W9-BUU-529

More advance praise for *Strategic Renaissance:*

"The first rule for any successful strategy is to think. In *Strategic Renaissance,* Evan Dudik inspires the reader to do just that. The book is an imaginative almanac of great ideas."

—Thomas C. Shull, CEO, Meridian Ventures, Inc., and former CEO, Barneys, New York

"*Strategic Renaissance* is a must-read for CEOs and others constructing strategies. The fundamental concepts of hammer, pivot, bearing, and hammerhead provide clarity for strategic formulation and communication. The '81 Do's and Don'ts on the Road to a Great Corporate Strategy' are a godsend in themselves."

—J. Thomas Williams, President and CEO, Thomas Group

"Strategy is an art that works best when supported by science, and *Strategic Renaissance* combines both in an intriguing and refreshing way. The book is a hurricane of ideas and examples—I could not put it down."

—Vladimir Merson, Managing Director, Deutsche Bank, Moscow

"A must for anyone interested in morphing their organization to higher levels of performance—for any company needing a strategic renaissance!"

—Ram Ramakrishnan, Vice President–Strategy and Development, Square D Corporation

STRATEGIC RENAISSANCE

New Thinking and
Innovative Tools to Create
Great Corporate Strategies . . .
Using Insights from History and Science

Evan M. Dudik

AMACOM
American Management Association
New York • Atlanta • Boston • Chicago • Kansas City • San Francisco • Washington, D.C.
Brussels • Mexico City • Tokyo • Toronto

Special discounts on bulk quantities of AMACOM books are available to corporations, professional associations, and other organizations. For details, contact Special Sales Department, AMACOM, a division of American Management Association, 1601 Broadway, New York, NY 10019.
Tel.: 212-903-8316. Fax: 212-903-8083.
Web site: www.amanet.org

Grateful acknowledgment is made for permission to reprint excerpts from Influence *by Robert Cialdini. Copyright © 1993 Robert B. Cialdini. Reprinted by permission of HarperCollins Publishers, Inc.*

Library of Congress Cataloging-in-Publication Data

Dudik, Evan Matthew.
Strategic renaissance : new thinking and innovative tools to create great corporate strategies—using insights from history and science / Evan M. Dudik.
p. cm.
Includes bibliographical references and index.
ISBN 0-8144-0551-7
1. Creative ability in business. 2. Discoveries in science—Case studies. 3. Industrial management—Philosophy. 4. Strategic planning. 5. Success in business. I. Title.
HD53.D83 2000
658.4'012—dc21 *99-057944*

Printing number

10 9 8 7 6 5 4 3 2 1

To Betty E. Dudik,
without optimism like hers, nothing would be created

And to George F. Dudik,
without perception like his, nothing would survive

Contents

Acknowledgments

One rainy, muddy Virginia day as a first-year graduate student, I borrowed my father's new car to deliver a paper I had detested writing to a professor I loathed. When I arrived at his country home reasonably close to the appointed time, I rang the doorbell, to no result. After several more rings, a woman in a pale blue nightgown appeared. Obviously, I'd roused the professor's wife from her sickbed. I handed over the paper and tromped back through the mud to the car. I backed the car down what I thought was the driveway but ended up demolishing the professor's mailbox and prominently denting the new car's trunk. Calling to apologize to the professor about the mailbox, I learned that the paper I'd delivered was critical of the professor's most esteemed mentor and closest personal friend.

When it comes to writing, it's been pretty much uphill from there. And that's primarily because in this book I've been able to write about the largely enjoyable experiences I've had as an academic, project manager, lobbyist, management consultant, and entrepreneur in the company of the world's finest clients, customers, and colleagues.

This book is based primarily on the experiences of my consulting clients and my own manufacturing entrepreneurial experiences. So my first debt is to my many clients whose triumphs, persistence, and lessons I've had the honor to share. Their life's work ranges from transportation to health care, from the Internet to wood products, from retailing to printing. There are too many to name in this short space, and it would be unfair to link their names to the interpretations in this volume. I'd like to acknowledge, however, early reviewers of this book's drafts who were generous with their time, liberal with their encouragement, and on target with their criticism: Phil Fiske, Alan Kritzer, Bob Neely, Mike Nevens, and Doug Smith. Editor Frances Caldwell labored to bring order to the first set of ideas and found the real-world correlates to citations buried in my memory. Editor Dan Wilson did not spare himself in examining prose and logic. No book survives without a champion. In this case it was AMACOM executive editor Adrienne Hickey, whose return from vacation was marred by the arrival of my manuscript.

I'd like to acknowledge the formulators of the intellectual and experiential background of this book. First are my colleagues and mentors at the consulting firm of McKinsey and Co., Inc., especially those in the Los Angeles office, who took a risk on inducting a former philosopher and lobbyist into the firm. Second are my colleagues at Meridian Ventures and R. B. Webber and Company—Paul Jen, Ed Lambert, Tom Shull, and John Shull of the former; Steve Jordan, Todd Kelleher, Robert Lauridsen, and Jeff Webber of the latter—who showed how fast acting and on target the best management consulting can be. Third are my professors and friends at Harvard Business School, especially the late Robert Abernathy, and Al Chandler, Bill Fruhan, David Garvin, John Kotter, and Howard Stevenson. Fourth and most fundamentally are the teachers at St. John's College in both Annapolis, Maryland, and Santa Fe, New Mexico, every one of whom has an unshakable belief in firsthand encounters with first-rate thinkers. I hope that to thank Eva Brann, Jacob Klein, and Richard Wiegle is also to acknowledge their contemporaries and successors for the intellectual, pedagogical, and, not least, managerial success they have achieved, often in the face of daunting odds. Finally, Edwin B. Allaire, Brian Cooney, Charles Hartshorne, Bob Kane, Irwin C. Lieb, David Miller, and Alex von Schoenborn of the University of Texas at Austin showed that uncommon intellect can be a common virtue in the right place at the right time.

And most importantly, Jane, Graham, Tristan, and Matthew provided the joy to life that makes every undertaking worthwhile.

Naturally, all the judgments, errors, and omissions in this book are mine alone. My only hope is that the small errors may be overlooked and that the large errors prove interesting.

Chapter One

What Strategy Has Learned from Astrology and What It Needs to Learn from Science

> What we have learned
> Is like a handful of earth
> What we have yet to learn
> Is like the whole world
>
> —AVVAIYAR

> Truth lives, in fact, for the most part on a credit system. Our thoughts and beliefs "pass," so long as nothing challenges them, just as bank notes pass so long as nobody refuses them. But this all points to direct face-to-face verification somewhere, without which the fabric of truth collapses like a financial system with no cash-basis whatever.
>
> —WILLIAM JAMES

Early on the morning of 17 February 1600, in Rome's Field of Flowers, after eight years of imprisonment and interrogation and more than a decade of wearing out his welcome in the courts of Europe's aristocrats, Italian mystic philosopher Giordano Bruno found himself naked and bound upright by stout ropes to a stake. Soon, acrid smoke and leaping flames rose high around him. Bruno, a fugitive, defrocked Dominican friar, was guilty of proclaiming in public three interesting ideas: first, that there was an infinite universe; second, that there not only might be but almost certainly were other inhabited Earths wandering in the skies above; and third, that Earth was a small planet revolving around a medium-size, or even small, star. Bruno was paying the ultimate price for his hand in pushing forward what later generations would call the Renaissance. And indeed, this theory of Bruno's was the rebirth of a thousand-year-old Greek speculation as to the size, composition, and inhabitation of the universe.[1]

Not ten years later, in late October 1609, the brilliant and caustic Italian philosopher and scientist Galileo Galilei peered from an upper-floor balcony of his villa into the night skies above the city of Padua, not far from Venice. He was squinting through a wooden tube of his own fabrication and design, though not of his primal invention. This telescope, invented barely a year before by a Dutch spectacle maker, resembled nothing so much as the crosspiece of an automobile lug wrench: a tube about two feet long and an inch wide, flared at each end to accommodate two tiny glass lenses. How Galileo saw was not as important as what he was the first human to see: that the moon was not a perfect, luminous sphere but had hills, valleys, and craters. He was the first to see that Jupiter had moons that revolved rapidly around it and not Earth. He observed that Venus had phases like Earth's own moon. And he saw what others considered impossible imperfections but what we now know are sunspots.[2]

Galileo, as we all know, did not pay as gruesome and final a penalty for his part in the Renaissance as did Bruno. But after twenty-two years of relative circumspection, he published a forceful defense of the heliocentric Copernican theory resting considerably on his astronomical observations and theories of physics. The Roman Catholic Church, after ample warning, summoned him to Rome and trial in 1632. His recantation of the Copernican system under duress at a church trial in 1633 saved his life but forced him into house arrest for the remainder of his life—something that may sound positively inviting to the road warriors reading this book. Nevertheless, by 1638, descending into blindness, he managed to complete his famous, cosmology-shattering book *Dialogue Concerning Two New Sciences.* Galileo smuggled it from his house to a printer in the more idea-tolerant Netherlands—a country that was fighting for its independence from both Catholic Spain and the Inquisition. It was, not coincidentally, Europe's rising financial, commercial, and intellectual powerhouse.

This is a book about corporate strategy. It is based on my belief that the time is ripe for corporate strategy, like science at the time of Galileo, to break out of the chains and dogmas of its own Middle Ages. Like the Renaissance itself, it is time for corporate strategy to reach back into the richness of human history and reach out to the habits, logic, and values of science.

Galileo and Bruno are apt heroes for this book because they exemplify some of the major themes I am inviting you to explore. First, Bruno was a bold inventor with far-reaching hypotheses of universal weight and grave significance. That should be a motto for any corporate strategy. No half thoughts for him—or for us!

Second, Bruno was certainly a radical, half thinker and half mystic, and something of an agent provocateur. His arrest, torture, and trial could have come as no real surprise to him. At least some of our strategic thoughts should be as provoking as his. Third, his ideas reached both back in time to ideas stowed in the vast library of human thought—in his case, to the Greek, Roman, and Islamic thinkers and observers—and so far forward in time that they reach even to the present day. So, too, should our thinking about strategy embrace what we know best about the nature of humanity and markets and competition from all ages through today as we rush headlong into an uncertain future. Finally, Bruno was perhaps the first Renaissance figure to espouse the idea that the progress of knowledge is limitless. So, too, should our thinking about what a company should do to continually evolve and even consume itself, rather than seek the safety of a single strategy that can stand the test of indefinite time.

For his part, Galileo exemplifies in other ways all the best about strategic thinking. He sought new facts about old phenomena—not just with the telescope but with his observations of the regular period of the pendulum, his investigation of falling bodies and hydrostatics, and the invention of the microscope. He settled not for the extensive, apparently cogent explanations offered by the scholars of his time and accepted for hundreds of years. He employed the best technology available to search for new facts—inventing some along the way. He repeated his observations until he could have every confidence in them.

This search after observable facts, not to verify but to imperil our strategic ideas, is the heartbeat of this book. Then, too, Galileo performed this search himself, personally, not subcontracting either the manufacture of his telescope or his observations or thinking, as is so often done by today's corporations. He did not stop there. He did not expect the facts to speak for themselves. Rather, he wove them into a comprehensive, logical, and experimental theory. Finally, even Galileo's trial, involuntary as it was, highlights the rigor we should expect from our strategic thinking and from ourselves.

It might seem shocking that today, in this high-tech age, I should be calling for an end to the Middle Ages in corporate strategy. That age was characterized not by scholarly sloth but by an almost frenetic argumentation that seems, looking back, headed nowhere. A survey of business strategic thinking makes a convincing case that we are in a kind of strategic Middle Ages. Let's take a quick look.

The Silver Bullet Market

Business leaders, executives, and strategists have looked for years to business books for a polestar for strategic success. And authors have been busy writing them. About 2,000 a year are published.[3] If there were a single strategic framework that worked, it surely would have been found and written up by now. If you've sought—and failed to find—guidance on the shelves of your favorite bookstore, you're not alone. My bookcases groan with volumes promising the secrets of successful strategy—with agendas that seem promising and productive—but in the testing lab of businesses I've been in and those of my clients large and small, they've turned out to be impractical, unprofitable, and incomplete.

It's understandable. There's more than a shred of truth in both the newest and the oldest business strategy concepts, and there's value in books that use dozens of examples. There's also truth in one-company-got-it-right books such as Andy Grove's paean to Intel's success, *Only the Paranoid Survive;* or *Sam Walton—Made in America: My Story;* or *Powered by Honda: Developing Excellence in the Global Enterprise.*[4]

Like the ubiquitous diet books of recent years, each of the strategy books claims that the others are outmoded or false. Yet each book typically contains dozens of stirring examples that purport to demonstrate the wisdom of its recommendations. The diet books, health magazines, and TV fitness commercials offer glowing testimonials from jubilant women who have lost fifty pounds or male models with washboard abdominals. The strategy books point to companies and corporations that have—according to the authors of the books—followed the strategy recommendations the books offer and are enormously successful *as a result of their strategy.*

And the books' strategies are as different from one another as the diets. In the late 1970s, strategy was all about searching for excellence by Sticking to Your Knitting. And Tom Peters made money. In the early 1980s, it was all about Quality. And Philip Crosby and Edwards Deming made money.[5] I loved Crosby's approach and became a convert. Then we Made the Customer Number One. And *everybody* made money.[6] The base premise was indisputable. Then it was Reengineer; this was roughly Replace People with New Processes and Information Systems. And Michael Hammer and James Champy and CSC and Gemini Consulting made money.[7] And yes, there's a lot to be said for reengineering instead of mere tinkering. But then came chaos theory from brilliant mathematicians and global climate experts. So we Throve on Chaos, and Peters made money again.[8] Then came the Wisdom of Teams,[9] a desperately needed antidote to America's hero-a-minute culture. Now, as of this writing, it's Create Your Business Network and Get on the Internet.[10] No doubt creating self-reinforcing networks *is* crucial. But sometimes it looks as though e-commerce strategy books are the most profitable part of the whole Internet business.

The jarring thing, of course, is that there is plenty of *truth* in every one of these strategy recommendations. How can you be against excellence or making the customer number one? Lots that's worthwhile. And that's just the problem: too much truth—too much *confirming evidence.*

Our Adventure Begins with Astrology

What does *astrology* have to do with business strategy? Unfortunately, everything. To see why, and what to do about it, we have to go on an intellectual excursion—in fact, this whole book is just such an adventure. Our purpose here is not to promote a new strategy (or a new diet). Instead, we're going to go into the heart of the concept of strategy. We're going to look at history and nature and see what each can teach us about strategy. We're going to look at history and nature and see what each can teach us about strategy. We're going to shed the straitjacket of business anecdotes and look at the logic of scientific discovery. We're going to look at what makes any good strategy work—the very *conditions* for strategy.

We're not going to be just conceptual; we're going to be practical. I promise that by the end of every chapter you will have discovered the possibility for doing something different tomorrow. But without a firm grasp on the conceptual foundations of strategy, we'll just flounder among the anecdotes of how Toyota *does* it, how Microsoft is *doing* it, and how Wal-Mart *did* it.

I said that the problem with the multitude of strategic approaches is that they supply too much truth—too much confirming *evidence.* Our excursion begins with why that's so—and why we need not the right way to be *right,* but the right way to be *wrong.* Ready?

A friend of mine, Professor Bob Kane, consented to appear on television to debate the validity of horoscopes with a professional astrologer. As an expert on the logic of science, Bob expected it to be a simple task to squash the astrologer like a bug on the windshield of scientific method.

To his amazement, Bob was thoroughly thrashed.

Unexpectedly, the astrologer had truckloads of supporting facts at his disposal. The astrologer could cite case after case in which the stars had predicted natural cataclysms, wars, and the deaths of famous personalities. He had *hundreds* of confirming examples. He just started reeling them off, hardly giving Bob a chance to get a word in edgewise.

The astrologer left the podium in triumph, having won the debate not on mysticism or appeal to the emotions but on the *empirical evidence.*

The astrologer, I'm afraid, was using the same basic logic as the strategy book writers. They provide lists of companies in which their specific strategies worked. Likewise, the astrologer provided lists of events in which astrological predictions were borne out. In both cases there are *heaps:* collections of large numbers of things that apparently worked in accord with the theory.

Question: What's wrong with these pictures?

Answer: Astrology is not proved by a catalog of examples, and a business strategy is not proved by a list of successful companies. It simply isn't that easy. You don't prove that the sun orbits the earth by pointing to all the sunrises in history. You don't prove that a diet works by dredging up hundreds or even *thousands* of ecstatic testimonials.

Writers compound the error. First, they manipulate the evidence, often unconsciously. One popular—and in many ways admirable—book, *Marketing Warfare,* a self-proclaimed strategy handbook that put advertising gurus Jack Trout and Al Ries on the power-audience lecture circuit, offers numerous strongly drawn insights into business strategy. But it stretches the evidence to fit the scheme. For example, Trout and Ries want to make the point that market-share leaders are always perceived as better. So, when they observe that Ford and Chrysler were actual leaders in product and technical innovation, they ask, "who gets the credit for engineering excellence?" and the answer is "General Motors, of course."[11] Yet by 1986, when *Marketing Warfare* appeared, Toyota, Nissan, Mercedes, and BMW were in the arena challenging all comers for "the credit for engineering excellence." GM was hardly in the running.

The famous *In Search of Excellence* does a different kind of dance to *prove* that its cornerstone advice, "Stick to your knitting!" is strategically sound. 3M is one of the book's hero companies. It is a company famous for innovation—with tens of thousands of products in dozens of distinct markets—and it insists that 30 percent of its sales is from products that didn't exist four years earlier. Frankly, this looks to me like a good example—perhaps even a *great* one—of *not* sticking to your knitting. However, Peters and Waterman, faced with the then-undeniable success of 3M in the 1970s and mid-1980s, make *innovation itself* the knitting.[12]

By the way, where is 3M today? It has new products such as pliable fiber-optic cables for electronic desktops and a viable replacement fluid for chlorofluorocarbons in air conditioners. So, it has innovation to offer, from telecommunications to refrigeration supply. It also has flat sales, a declining return on invested capital, molasses-like inventory turns of 3.5, and a bloated headcount.[13] If *In Search of Excellence* were written today, Peters and Waterman might just use 3M as an example of what happens when you *don't* stick to your knitting.

Remember huge conglomerates such as ITT and Grace and Textron? They are *Excellence*'s natural examples of not sticking to your knitting. How would the conglomerators reply to that charge? Easily. Boosting earnings per share through financial engineering and centralized overhead was what they specialized in. In fact, that's exactly what conglomerators like Peter Grace and Harold Geneen would tell you their knitting was.

It's hard not to distort the evidence when you have an exciting idea. It's just human nature. And the practice isn't limited to businesspeople, by any means. When eminent American astronomer Percival Lowell first observed Mars in the 1890s through the highest-powered telescope then available, he was so certain he saw canals created by living beings that he went to the press with the news. He published *Mars and Its Canals* in 1906.[14]

In 1989, Stanley Pons and Martin Fleischmann announced the existence of cold fusion. These are professionals who were at the top of their fields—true rocket scientists. Fleischmann is a fellow of the Royal Society; Pons is a former chair of the department of chemistry at the University of Utah. Most scientists now believe that they either misinterpreted their data or contaminated their apparatus, although a few diehards still think cold fusion exists.

Another way the strategy books make astrology-like errors is almost the opposite of creating a framework and looking for confirming examples. It's picking a successful company and then attempting to find out *what makes it tick.* The underlying logic goes this way: Behind every successful company *must be* a successful strategy. Let's search out what successful companies have done, and voila! The secrets of strategic success. Strategic writing has involved collection after collection of examples without end, seizing on successful companies, reverse-engineering their success, and pronouncing the results *the* strategies for the 1970s, the 1980s, the 1990s, or, for that matter, all time.

Is there a better way? A better way than searching for strategy by example?

Blame It on Bacon

What we have here is the revered *empirical method* run amok. Let's go back in history about 400 years, to England's Sir Francis Bacon (1561–1626). As a rule, today we expect politicians and judges like Bacon who plead guilty to bribery to drop from sight, undergo religious conversion, or become lobbyists. Not so Sir Francis. While the leaky *Mayflower* was averaging two miles per hour in its storm-tossed crossing to North America,[15] Bacon was trying his hand at philosophy. His *Novum Or-*

ganum; or, *Indications Respecting the Interpretation of Nature* (1620) cleared the pathway for modern scientific thought.

Bacon was one of those Enlightenment thinkers who started the West pecking away at the shell of medieval Scholastic thought. To oversimplify, medieval Scholastics tried to arrive at truths about everything—from the flight of an arrow to the morality of war—by relying on logical deductions from firm *first principles.* The idea was that if the first principles are firm and the logic is firm, the conclusions must be firm. Not a bad way to go, it would seem. Mathematics created its own universe of thought on just this foundation. It was spectacularly successful. Thus from first principles, truths about nature and God and human conduct could be deduced—or so it was thought—by logical enough, studious enough, clever enough thinking.

Unfortunately, there was a dearth of agreed-upon first principles. In fact, people disagreed more often than they agreed about them—regularly starting wars and burning one another alive over the *true* first principles. In fact, the Church had to go through considerable intellectual gyrations to figure out exactly what Bruno's crime was. And for years, the Church was willing to let Galileo get away with supporting Copernicus's heliocentrism as long as it was spun not as literal truth but as an elegant mathematical scheme for predicting celestial events.

Scientists, philosophers, and mathematicians struggled to find a way out. Bacon's contemporary René Descartes was content with finding *just one* indubitable first principle. At least, it seemed firm to him. A cold winter evening by the wood stove in a hut in Holland produced his minimalist "I think, therefore I am." But even then, Descartes had a difficult time proving from that apparently firm foundation that the *real world* existed. It was a scandal from which philosophy has yet to fully recover.

Faced with all this (and with plenty of time on his hands), Sir Francis was among the first to (re)introduce the concept of induction to ears that were finally ready to hear the message. Instead of searching for firm first principles, Bacon proposed a huge program of gathering empirical information. His idea was to base the formulation of general truths on the exploration of many particulars. He turned the scientific world upside down, placing induction from experience at the center of the quest for knowledge about nature. Little did he know, as a gentleman and no tradesman, that he had set the stage—and dug the hole—for four centuries of business strategy.

In one particularly vivid case, Sir Francis (who apparently never got his hands dirty doing actual scientific work) recommended a better way to understand *heat.* He suggested gathering examples of the various kinds of heat. The scientist should gather flaming brands, glowing coals, heat from peat bogs, the heat of animals, heat from mixing chemicals. These different examples of heat should be examined meticulously and all their commonalities and differences noted. Then all the samples of heat could be categorized into heaps—heat heaps. Each time the Baconian scientist was presented with a new example of heat, he could examine it and assign it to the proper heap. Using his prior observation of the old examples in that heap, he would be able to predict (or so the idea goes) what this new example would do.

This was surely an advance over late Scholastic logical deductions of what heat *must* be from its *nature* (although the cleverness and insight of Scholastics should not be discounted). But not all was brilliance and happiness in the house of Baconian induction. Scientists who *did* get their hands dirty—and philosophers who *didn't*—soon realized that this sort of bootstrap science had severe, debilitating limitations.

For one thing, it had no good account of cause and effect. In fact, this approach *discounted* cause and effect in an attempt to avoid medieval Scholastic speculation about what the causes, the effects, and especially the *purposes* of things and events *must* be, based on logic: the nature of things, according to Scripture and Aristotle. Bacon desperately wanted to rid science of hidden causes. He wanted desperately to link together those things that the human senses could perceive.

Then, too, Bacon seems to have downplayed the role of quantification—the amount of heat of each kind. Perhaps this was because reliable instrumentation and reasonable standards hadn't yet evolved.

But there was an even more fundamental problem at the heart of the heap-making. Establishing the *criteria* for categorization (of heat heaps or any other kinds of heaps) is a *very* tricky business. Take a pig whose heat you are trying to categorize. Is the pig's breed a significant characteristic? Is its weight relevant? Is the intensity of the heat fundamental data? Is the presence or absence of flame important? Do differences between pig-generated heat and sun-generated heat have to be taken into account? Does the time of day matter? If any of these are in doubt, then the categorization of this heat as the same as or different from other samples, and the prediction that it will behave as other samples in its heap do, is in doubt. In short, the myriad particularities about each sample make the Baconian project a good place to start, but no place to draw definitive conclusions.

This is the fundamental problem of business strategy. There are lots of heaps, lots of great thought-starters, lots of tantalizing generalizations and seductive testimonials, and lots of *dead ends* for any particular company—especially yours.

No wonder businesspeople complain incessantly about making the predictable journey from enthusiasm through frustration to chagrin and ultimate disappointment with each new management fad. We think there's *got to be* a pony in there somewhere. And we keep searching hopefully through the heaps of success examples.

Like the astrologers, we find confirmation of past predictions everywhere in the historical record. We have a habit of driving by looking in the rearview mirror. But when it comes to predicting the future—our stations in life or business, the durability of our marriages, the performance of our investments—well, the record isn't so clear.

If you think this is all too theoretical, then you weren't an investor in Long-Term Capital Management (LTCM) during its spectacular September 1998 meltdown. Never was there a more beautiful example of driving—betting big money—by looking in the rearview mirror and predicting from Baconian heaps. Unfortunately for its investors, LTCM lost billions in capital during 1998, dropping from $4.8 billion at the year's beginning to $600 million nine months later. Let's look at what LTCM did.

LTCM was a huge securities hedge fund. It didn't bet on whether securities markets would rise or fall. What it bet on was that yields (and therefore prices) of comparable securities would converge, aligning by predictable amounts until they were fairly valued. John Meriwether, LTCM's searingly brilliant chairman, had reams of statistical data showing that throughout history, various securities had converged in price as rational investors came to regard them as of essentially the same risk, value, and quality. LTCM would pounce if two securities differed more than their history predicted, hoping to make a profit as other investors realized their *intrinsic worth* and bid one or the other security up or down.

For example, in June 1998, a 29-year treasury bond yielded $0.50 per year more than a 30-year treasury bond yielded.[16] Rationally—*scholastically*—the prices of these two securities should converge because they are so similar. They did so historically, according to statistical research—and according to Meriwether's personal experience trading bonds during the New York City bankruptcy threat in the mid-1970s. Lots of very bright people agreed with him and invested in or lent to LTCM: two Nobel Prize winners, Citicorp, CS First Boston, Switzerland's UBS AG, and the chairman of Merrill, Lynch.

Now, $0.50 a year is not much to buy lunch with, so LTCM borrowed heavily. By hedging these two securities (buying one and selling the other short), LTCM stood to make a paltry $5,000 on a $1 million investment. However, as *Forbes* columnist Robert Lenzner points out, if you can borrow $1 million for $10,000 in cash, that $5,000 becomes a 50 percent return on your actual $10,000 cash outlay. Not bad.

By 1998, Lenzner calculates, LTCM's $4.8 billion controlled $160 billion in stocks and bonds plus associated derivatives worth a notional $1 trillion, for leverage of 33:1 or, nominally, 240:1.

But the future wasn't like the past. The Asian economies shrank, and their financial markets plummeted. Russia defaulted on its foreign debts. Buyers of securities got scared and stopped buying—''liquidity dried up.'' Securities that were supposed to converge didn't. To Meriwether's mind, the markets behaved *irrationally.* To cover its positions and pay its debts, LTCM had to sell its securities. But with fewer buyers, there was no one for LTCM to sell to: no market for the billions and billions LTCM had to dump because of its dependence on leverage and tiny spreads. With fewer buyers, prices had dropped so far that LTCM took a $4.2 billion bloodbath.

Looking back, we—and Meriwether—can see *why* those securities that historical analysis threw into the *overvalued* heap *weren't,* and those in the *undervalued* heap weren't either. All that ''confirming evidence'' cost LTCM's investors a bundle and Meriwether his reputation.

The fiasco proved once again, however, that if you're a big enough debtor you can twist the tail of the world with impunity. Federal Reserve chairman and financial market godfather Alan Greenspan, wary of what the total impact on world financial markets would be if LTCM actually went bankrupt, ''arranged'' for banks and brokerages to cough up $3.6 billion in new investments in a deal they couldn't refuse.[17]

But your company may not have the chairman of the Federal Reserve for a sugar daddy.

Sir Karl Popper Had It Right

We need the help of Sir Karl Popper, an Austro-English philosopher of science. A native Viennese, Sir Karl was knighted by Queen Elizabeth in 1965. He was a friend and theoretical sparring partner of Niels Bohr, Werner Heisenberg, and Albert Einstein, among many others. He was a world-class pianist. He is considered by many to be the last in the line of rationalist philosophers stretching back through Kant, Hume, Locke, and Aquinas to Aristotle (although he wouldn't have been in business if he hadn't challenged them mightily). He died in 1994.

Sir Karl developed a new approach to understanding what makes a theory great. It is not the amount of evidence that can be marshaled to *support* a theory—such as examples of life happening according to horoscopes, business surging according to prescribed strategy, and weight dropping because of diet. On the contrary, patterns emerge in *everything* we observe. The stars in the night sky reveal patterns. Cloud formations take on animal shapes. Indistinct patterns on the image of Mars become canals. Pattern recognition, a trait we often seek out in our best and brightest, comes cheap. A pattern, by itself, means nothing.

And once you have seen the shapes in the clouds or the patterns in the stars, it's hard—perhaps impossible—*not* to see them. They reveal themselves so strongly that once the Big Dipper becomes familiar, it's hard to see those stars as anything else. Think of friends you had in college who were convinced Marxists or Freudians or liberals or conservatives. They seemed to see proof of their theories everywhere. Popper wrote, in 1919:

> I found those of my friends who were admirers of Marx, Freud, and Adler, were impressed by a number of points common to these theories and especially by their apparent *explanatory power*. . . . The study of any of them seemed to have the effect of an intellectual conversion or revelation. . . . Once your eyes were thus opened you saw confirming instances everywhere: the world was full of verifications of the theory. Whatever happened always confirmed it. Thus its truth appeared manifest; and unbelievers were clearly people who did not want to see the manifest truth; who refused to see it.[18]

Popper was the first to see the solution to the question, How many verifications or confirmations does it take to make a theory true? He perceived that the whole question was backward. Popper saw that a theory or hypothesis cannot be proved by the examples that illustrate it. *A theory cannot be proved at all.*

Instead of certainty and incontestable proof, we have to settle for what makes a theory *great*. And that is something that sounds like the opposite of proof. A theory's power and greatness lie in its ability to specify what observations or consequences would make it false. Popper called this a theory's *falsifiability*. Thus, the test of greatness for a theory is *great explanatory power* coupled with *great specificity* about what observations in the real world would make the theory false.

That's why I said that a theory of business strategy should talk about being *wrong* in the right way.

What, exactly, does it mean for a theory to be *falsifiable*?

Take, for example, the famous key number of quantum physics, Planck's constant. This number, 6.63×10^{-34}, is plucked from all the universe of numbers to describe the relationship between the frequency of an oscillator (such as atoms with light shining on them) and the energy emitted by that oscillator. The basic formula, $E = h\nu$, states that the energy emitted has a minimum of Planck's constant, h, times ν, the frequency of the system. Quantum theory further states that energy comes only in whole-number multiples of this minimum, *E*.

What's remarkable about this number is its great specificity. Planck dares you to come up with an observation that shows that the relationship is 6.64×10^{-34} or 6.63×10^{-24}. Popper thinks this theory is great because it is highly specific and a precise prediction of energy emitted in the photoelectric effect you can study in the laboratory—an effect used by ordinary solar-powered calculators. The equation is very clear—perfectly precise—about what would count as falsification. It dares you to measure the energy yourself.

Let's look at some more examples, as Popper would have, before we apply this to business. Popper's scientific hero was Einstein. He notes: "Einstein's gravitational theory had led to the result that light must be attracted by heavy bodies (such as the sun). . . . As a consequence it could be calculated light from a distant fixed star . . . would seem to be slightly shifted away from the sun."[19] During an eclipse, you can take two photographs of the fixed star and observe the Einsteinian shift. Then Popper says: "Now the impressive thing about this case is the risk involved in a prediction of this kind. If observation shows that the predicted effect is definitely absent, then the theory is simply refuted."[20]

How does this approach to a theory's greatness affect such ideas as astrology, Marxism, and psychoanalysis?

Astrology doesn't pass muster. Astrologers are greatly impressed, and misled, by what they believe to be confirming evidence—so much so that they are quite unimpressed by any unfavorable evidence. Moreover, by making their interpretations and prophecies sufficiently vague, they are able to explain away anything that might be a refutation of the theory. According to Popper, "The Marxist theory of history . . . ultimately adapted this soothsaying practice. In some of its earlier formulations, the predictions were testable, and in fact falsified. Yet instead of accepting the refutations, the followers of Marx re-interpreted both the theory and the evidence in order to make them agree."[21]

Surely nearly everyone reading this book has had the experience of dealing with a fanatical Freudian. Popper's analysis: "The two psycho-analytic theories [Freud's and Alfred Adler's] were in a different class. They were simply non-testable, irrefutable. There was no conceivable human behavior which could contradict them. . . . Those 'clinical observations' which analysts naively believe confirm their theory cannot do this any more than the daily confirmations which astrologers find."[22] Take for example, Popper says, the two cases of

> a man who pushes a child into the water with the intention of drowning it; and that of a man who sacrifices his life in

> an attempt to save the child. Each of these two cases can be explained with equal ease in Freudian and in Adlerian terms. According to Freud the first man suffered from repression . . . while the second man had achieved sublimation. According to Adler the first man suffered from feelings of inferiority (producing perhaps the need to prove to himself that he dared to commit some crime), and so did the second man (whose need was to prove to himself that he dared to rescue the child). I could not think of any human behavior that could not be interpreted in terms of either theory. It was precisely this fact—that they always fitted, that they were always confirmed—which in the eyes of their admirers constituted the strongest argument in favor of these theories. It began to dawn on me that this apparent strength was in fact their weakness.[23]

In November 1989, Marxism was falsified for the final convincing time with shovels and pickaxes in Berlin, Germany. In January 1997, Dr. Alan Stone, former president of the American Psychiatric Association and a once-convinced Freudian, asked from the cover of *Harvard Magazine*, "Psychoanalysis failed as science. Will it survive as art?"

Putting Popper to Work

What is a business strategy? It should be nothing more than a prediction about what markets, customers, and competition will do in the future, and how that will change depending on the action you take. That is, a strategy is a hypothesis. It is an "if-then" statement in which the *if clause* sets out the conditions and the *then clause* sets out the expected results precisely (see Figure 1-1). The more precisely the conditions and results are stated, the more testable and falsifiable the strategy.

The realization that a strategy is a testable, falsifiable hypothesis has great practical significance for you and me. First, vague strategies such as "consumers will like our lower prices" don't cut the Popperian mustard. A hypothesis must be more like, "if we can reduce average prices below competition by 3 percent, long term, our market share will grow ten points." If you're in the superstition business, a falsifiable hypothesis is that "on Friday the thirteenth, on average at least 5 percent more people die from natural disasters than on other days of the week."

Looking at business strategy as a falsifiable hypothesis has cash-value outcomes for strategists and their companies.

Figure 1-1. Falsifiable strategic hypotheses have two main elements.

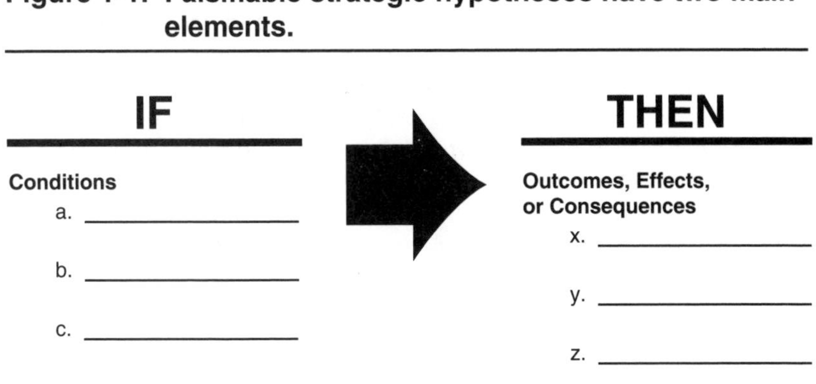

A Tale of Two Theories

Lately the business press has reported how Wal-Mart has been taking retail discount store market share from Kmart. The last ten years have been a tale of two strategies. Kmart has focused on upgrading its merchandise and store atmosphere with a veneer that isn't *quite* Sak's Fifth Avenue but tries. Wal-Mart has focused on reducing logistics costs through sophisticated inventory management, improving vendor relations, and cutting transportation costs. Wal-Mart's cost-cutting drive has proved to be more significant long term than Kmart's makeover. Everybody likes a winner, so the press has been glorying in the cleverness of the Wal-Mart strategy.

But what is important right now is not who is winning and who is losing. What is significant is that each company was betting on a testable hypothesis. Wal-Mart was betting that lower overall prices would gain long-term market share. Kmart was betting that demographics and consumer psychology required a store and merchandise upgrade.

Now Wal-Mart reportedly makes an investment of half a billion dollars a year in information technology. It has a database second in size only to that of the U.S. Government and several times larger than that of the next largest retailer, Sears.[24] This information is used to track store-by-store consumer buying patterns and vendor prices (almost instantly matching competitors' sales) and to lower transportation, warehousing, and inventory costs. Wal-Mart planned to trim inventory alone by $1.5 billion in 1997–98.[25] That's a big bet that, long term, somewhat lower prices will be the deciding factor in gaining market share. Yet it is a testable hypothesis.

Wal-Mart could have selected a dozen stores and lowered prices slightly in each, then tested customers' reactions. Likewise, Kmart could have upgraded a few stores in sample markets and checked out its hypothesis. Wal-Mart would have been simulating the price savings it could expect to pass on through lower logistics costs—and, by the way, determining how much lower those logistics costs would have to be to make a difference in the marketplace. Both companies had the opportunity to generate an "*if-then*" hypothesis, test it in a fraction of their stores, and then expand the process or drop it entirely, depending on results.

What might their *if-then* statements—their testable strategies—have been?

Wal-Mart: *If* prices are lowered storewide by an average of 2.5 percent, *then* market share in trading areas where we compete with Kmart will increase by at least 5 percent compared with stores without price reductions over a period of twelve months. *Further*, gross margins can be retained at current average levels because logistics cost reductions equaling the lost gross margin can be achieved.

Kmart: *If* store atmosphere is upgraded through an investment of $200 per square foot per store on the model of Target department stores, selection is upgraded by an average cost per item of 2 percent, and upscale advertising focusing on value and selection worth $50 million per year market share is launched, *then* in markets where we compete with Wal-Mart and Target, market share will increase by at least five points, and average gross margin will increase at least 10 percent. Take a look at Figure 1-2.

Do successful companies do this? It doesn't confirm our idea (Popper would say), but in fall 1998, Wal-Mart announced that it was building giant "experimental" grocery stores to see whether its discount retailing concept could be extended into the juicy, $450 billion grocery market.[26] And there are indications that it will bring the same

Figure 1-2. Hypothetical Kmart strategic statement illustrates precision, specificity, falsifiability.

IF		THEN
a. Store atmosphere is upgraded by $200 per sq. ft. . . .		In markets where we compete with Wal-Mart and Target . . .
b. Product selection is upgraded by an average of 2% . . .		a. Market share increases by at least 5 points
c. Advertising is refocused on new value, selection with additional budget of $50 million . . .		b. Average gross margin increases by at least 10% (e.g., from 35% to 38.5%)

dedication to low everyday costs to a seemingly unrelated business: consumer banking.

The Plan for the Remainder of This Book

If you've read this far, you are already far ahead of most corporate strategic thinkers in companies today. But applying these insights about falsification versus verification to the real world is a journey that will take us through a three-stage adventure.

In chapters 2 and 3 we'll recognize that having a strong, falsifiable hypothesis isn't enough. It turns out that great strategies have important structures in common. So we'll perform some strategic dissection, using military history and other tools to make sure that your strategic body has all the necessary parts.

In chapters 4 and 5 we'll see why one well-known touchstone for a great strategy—whether it creates a Sustainable Competitive Advantage (SCA)—is powerful but desperately obsolete. However, SCA gives way to a new vision of strategy—one that recognizes that strategy has a life cycle, like other living things. This is OCE: Opportunity Creation and Exploitation.

Chapters 6, 7, and 8 make the most of the OCE insight. Chapter 6 provides concrete tools for better hypothesis generation and testing. It examines why market research and consultants are normally misused and what to do about it. Chapter 7 maps management's choices for making a great strategy happen in the real world. It sets out the rewards and dangers of different methods of creating strategic breakthroughs and exploiting those breakthroughs. Chapter 8 shows how OCE requires a radical new way of thinking about allocating, accounting for, and budgeting resources. One or two of these ideas will drive Wall Street crazy but give managers much more control and flexibility.

Chapters 9 and 10 deal with the human side of strategic success. In chapter 9 we explore why success's biggest enemy is the invisible, gravitational force known as culture. But we're not content with that surface insight. Instead, we dive deeper and look at the causes and life cycle of corporate culture itself and its precise impact on strategy. In chapter 10 we again gather insights from history and psychology to provide tools for your company's top management team to reshape culture to strategic needs.

Finally, chapter 11 summarizes in 81 do's and don'ts just what companies should do to create, test, and implement winning corporate strategies. You may be tempted to tear out these pages.

That's the plan of the book. For the moment, let's return to the fertile topic of strategic hypothesis making and falsification.

Practical Consequences for You to Take Away

I promised you that every chapter of this book would provide ways in which you might approach your strategic job differently. Even though we've just begun, you've learned enough to make a difference tomorrow.

1. The Kmart–Wal-Mart example shows in bold relief that your company's strategic theory should have precise, definite, falsifiable features. If your company's current strategy doesn't have them, it's time to bring it into the garage for an overhaul.

2. Every strategy sets out the *conditions* and then the *consequences* you expect. No hypothesis says that *x* will follow *y unconditionally*. Does your strategy say what you expect to happen *if* you carry out certain actions? *If* the market or demographics are a certain way? *If* competitors introduce (or don't introduce) specified products or services? "Being number one or number two in every market we compete in" is a *mission*, not a strategy for accomplishing one. (See chapter 2 for more on this.)

3. Similarly, every strategy should invite—even dare—you to say how it would be tested.

4. Nearly every strategy's *if-then* statement can be modeled. We don't necessarily need to kill the patient in order to find out whether the medicine works. Wal-Mart could have tested in a dozen locations by simulating logistics savings with low prices. Kmart could have tried its upscale strategy by investing in a dozen locations (and maybe it did this). There are numerous sophisticated research and survey techniques that can simulate shopping experiences for both consumer and industrial sales. With a little imagination, they can become competitive games that permit you to model how competitors will react and how markets and prices will change in the real, dynamic world.

5. When experience or experimentation falsifies a theory, you've *learned* something. Perhaps the original idea was close. Perhaps the original idea worked under some conditions but not others. The practical implication? Strategy is a work of constant refinement. Check your company's strategy. Does it allow for refinements? Is it all-or-nothing? Netscape survived at high valuations for five years because it refined and shifted its strategy. If it had stayed with its original blend—focusing on bringing the Internet into every home—there would have been nothing left for America Online to buy when it announced its blockbuster purchase of Netscape in December 1998.

6. Don't judge—don't spurn—a strategic theory because of its origin. Johannes Kepler hypothesized that planets' orbits traced ellipses instead of circles around the sun. This later became crucial for Sir Isaac

Newton's theory of gravitation and the invention of the calculus. But Kepler's theory was generated partly because of his religious beliefs about astronomy and the relationship of heavenly bodies to music. Creativity gurus like to tell the story of Friedrich von Kekulé, who fell asleep while watching the dance of flames in a fire. He dreamed of a snake that coiled and coiled around, finally swallowing its own tail. The result was the fundamental concept of organic chemistry, the carbon-based benzene ring. I'm sure that origin wasn't stressed in his scientific papers. Theories can come from anywhere. Their pedigrees (or the pedigrees of the consultants or executives who propose them) are almost irrelevant. What's particularly dangerous is the new executive or consultant who says that his or her idea has worked in dozens of other companies. Please award that person your star-speckled Merlin cap.

7. Likewise, your admonition to strategy consultants should be, "Once you've generated your recommended theory, I want to see every shred of evidence *against* that theory." Better yet, hire somebody whose sole job is to critique it, find evidence against it, falsify it. That'll keep them honest.

I know. *I've been on the receiving end.*

We now have some of the substance of what we should expect from sound business strategy. It sounds like a lot to demand of our companies—and ourselves. Fortunately, strategic theory tells us that often *less is more.* That's the theme of chapter 2.

Notes

1. William Boulting, *Giordano Bruno: His Life, Thought, and Martyrdom* (London: Kegan Paul, Trench, Truebner and Co. 1914); *Collier's Encyclopedia,* s.v. On-line Edition.
2. Sister Maria Clesete, *The Private Life of Galileo* (Boston: Nichols and Noyes, 1870); Albert diConzio, *Galileo: His Science and Significance for the Future of Man* (Portsmouth, NH: ADASI, 1996); Richard Panek, *Seeing and Believing: How the Telescope Opened Our Eyes and Minds to the Heavens* (New York: Viking, 1998).
3. *Forbes,* UK edition, 18 November 1996.
4. Andrew Grove, *Only the Paranoid Survive* (New York: Doubleday, 1996); Sam Walton and John Huey, *Sam Walton—Made in America: My Story* (New York: Bantam, 1993); Dave Nelson, *Powered by Honda: Developing Excellence in the Global Enterprise* (New York: John Wiley and Sons, 1998).
5. See, for example, Philip B. Crosby, *Quality Is Free* (Mentor Books,

1992); Philip B. Crosby, *Quality without Tears* (New York: McGraw-Hill, 1995); Philip B. Crosby, *Quality Is Still Free* (New York: McGraw-Hill, 1996); W. Edwards Deming, *Out of the Crisis* (Cambridge: MIT Press, 1986); Joseph M. Juran, *Juran on Leadership for Quality* (New York: Free Press, 1989).

6. Among dozens of titles, Robin L. Lawton's *Creating a Customer-Centered Culture: Leadership in Quality, Innovation, and Speed* (Milwaukee: American Society for Quality, 1993) is typical, if not above average.
7. Michael Hammer and James Champy, *Reengineering the Corporation: A Manifesto for Business Revolution* (New York: Harper Business, 1994); Michael Hammer, *Beyond Reengineering: How the Processed-Centered Organization Is Changing Our Work and Our Lives* (New York: HarperCollins, 1997); Michael Hammer and Steven A. Stanton, *The Reengineering Revolution: A Handbook* (New York: Harper Business, 1995).
8. Tom Peters, *Thriving on Chaos* (New York: Knopf, 1987; reprint, New York: HarperCollins, 1991); Tom Peters, *Embracing Chaos* (audiocassette, Nightingale-Contact, 1993); Tom Peters, *The Tom Peters Seminar: Crazy Times Call for Crazy Organizations* (Vingate, 1994); Tom Peters, *Liberation Management: Necessary Disorganization for the Nanosecond Nineties* (New York: Fawcett, 1994).
9. Teamwork is the critical factor in making a theoretical construct a strategy—a reality, as covered in chapter 10. A partial list of the outpouring on teamwork follows: Jon Katzenbach and Douglas Smith, *The Wisdom of Teams: Creating the High Performance Organization* (New York: Harper Business, 1994); Jon Katzenbach, *Teams at the Top* (Boston: Harvard Business School Press, 1997); Richard Willins and William Byham, *Inside Teams: How Twenty World-Class Organizations Are Winning through Teamwork* (San Francisco: Jossey-Bass, 1996); Pat Williams and James Denney, *The Magic of Teams* (Nashville, TN: Thomas Nelson, 1997); Robert Blake et al., *Spectacular Teamwork* (New York: John Wiley, 1987).
10. For example, John Hagel and Arthur Armstrong, *Net Gain* (Boston: Harvard Business School Press, 1997), has put Hagel on the speakers' circuit; Chuck Martin, *Net Future: The 7 Cybertrends that Will Drive Your Business, Create New Wealth, and Define Your Future* (New York: McGraw-Hill, 1998).
11. Al Ries and John Trout, *Marketing Warfare* (New York: New American Library, 1986), 59.
12. Thomas Peters and Robert Waterman, Jr., *In Search of Excellence* (New York: Harper and Row, 1982), 224–34.
13. *Forbes*, 19 October 1998, 54.
14. *Columbia Encyclopedia*, 5th ed., s.v.

15. James Daugherty, *The Landing of the Pilgrims* (New York: Random House, 1950), 43.
16. Robert Lenzer, "Archimedes on Wall Street," *Forbes,* 19 October 1998.
17. Ibid.; Gretchen Morgenson, "John Meriwether: Hedge Fund Wizard or Wall St. Gambler Run Amok," *New York Times,* 2 October 1998.
18. Karl R. Popper, *Conjectures and Refutations: The Growth of Scientific Knowledge* (New York: Harper Torchbooks, 1965), 34–45.
19. Ibid., 36.
20. Ibid.
21. Ibid., 37.
22. Ibid., 37–38.
23. Ibid., 35.
24. Christopher Palmeri, "Believe in Yourself, Believe in the Merchandise," *Forbes,* 8 September 1997; Emily Nelson, "Why Wal-Mart Sings, 'Yes, We Have Bananas!' " *Wall Street Journal,* 6 October 1998.
25. Nelson D. Schwartz, "Why Wall Street's Buying Wal-Mart Again," *Fortune,* 16 February 1998.
26. "Look Out, Supermarkets—Wal-Mart Is Hungry," *Business Week,* 14 September 1998; "Wal-Mart to Build Supermarket in Bid to Boost Grocery Industry Share," *Wall Street Journal,* 10 June 1998.

Chapter Two

"Only Make the Right Wing Strong": The Four Key Elements of a Successful Strategy

It must come to a fight. Only make the right wing strong.
—COUNT ALFRED VON SCHLIEFFEN, CHIEF OF THE GERMAN GENERAL STAFF, 1913[1]

Necessary but Insufficient

If you accept the idea that a robust strategy is a falsifiable hypothesis, you will have taken the quantum leap that moves a tough-minded company's thinking ahead of the pack. But the questions fairly clamor to be asked: How do we go about making a falsifiable hypothesis? What are the crucial elements of a strategy? How do we know when we have a strategic concept worth pursuing?

It's easy to see that the falsifiability of a strategic hypothesis is a *necessary* condition for creating a great corporate strategy but not a *sufficient* one. And there is no cookbook that lists all the remaining ingredients of a winning strategy, for reasons we'll explore in a moment. But what we can supply—and what this chapter is about—are the elements that history and logic suggest successful strategies must have. We can make these elements a useful yardstick for evaluating a strategic hypothesis you articulate, before you go to the trouble and cost of attempting to falsify it. And one of the crucial elements is the need to identify your company's right wing and make it strong.

But first it's important to understand why these guidelines or yardsticks are only that. What distinguishes our approach to strategy from others is an emphasis on the fact that a great strategy is an act of creation. It's important to everything that comes later to understand why there can be only guidelines to the elements of strategy and why—both as a matter of practice and in principle—there is not now and never can be any cookbook, any royal road to strategy creation.

I make a point of this because it is oh-so-tempting to grab a *trend* and call it a strategy, or to write up the ever-popular Strengths, Weaknesses, Opportunities, and Threats analysis (a worthy exercise in itself), find a niche (or shoehorn your company into one), and call the completed exercise a strategy.[2] But strategy can't be created by following a cookbook or using a software program, even if your business is a corner ice cream shop.

An Act of Creation

For starters, suppose there *were* a cookbook. There are indeed software programs that promise you a corporate strategy if you'll "just fill out these ten forms." First, if these programs were of value, it's obvious that your competitors would have access to them, too. And their reactions would be the same as yours: How do we beat the cookbook strategy? Suppose your strategy focuses on quality and reliability, like that of Federal Express. "We'll go you one better," says UPS. "We'll provide fair quality and adequate reliability and undercut prices by 10 percent. After all, we've already got 100,000 brown trucks on the street." This is an act of creation. Human creativity is inherently unlimited, and it will outfox a cookbook every time.

Second, the instructions don't guarantee the result. If there were ever a place for cookbook hypothesis creation, it is in the solving of those algebraic equations we all either loved or hated in eighth grade. After all, the rules are clear, there aren't many of them, and the equations aren't trying to outwit you at each step. Yet even here, the most a teacher can provide are guidelines to simplify the process of solving the equations. Some kids have a knack for choosing the right problem-solving strategy and getting to the right answer in just a few steps. Others use more steps and a number of approaches before one clicks. And still others may never find a problem-solving strategy at all but spin out perfectly valid hypotheses that lead back—with the mathematical ineluctability of circular reasoning—to the starting point.[3]

From five short postulates and five common notions, Euclid could prove not only the famous Pythagorean theorem (the square of the hypotenuse of a right triangle is equal to the sum of the squares of the other two sides; Proposition I.47) but also that equiangular parallelograms have to one another the ratio (in area) compounded of the ratios of their sides (Proposition VI.28) and that if a parallelepipedal solid is cut by a plane that is parallel to the opposite planes, then as the ratio of the base of one is to the base of the other, so will the volume of the solid be to the volume of the other solid (Proposition VI.25).

And in science, even the simplest hypothesis is a creative act. No mass of observations—no volume of data—will by itself generate a hypothesis worth testing. The great discoveries are acts of synthesis in which disparate ideas are combined and transformed to create something surprising and new. But not just any commingling of great ideas works. In science, the rewards go to the researcher with the insight to put together an idea that hits the mark—from the myriad ideas possible. Let's look at one.

Cholera epidemics in London in the 1840s and 1850s took thousands of lives. Nothing was then known of cholera's bacterial etiology. Death came from dehydration after severe diarrhea and vomiting. John Snow, a London physician, had the idea of charting outbreaks on a map of the city. After considerable effort and risk, Snow saw patterns emerge from the map. He noticed—as no one had before—that outbreaks clustered around common wells. From this, it was a short step to the idea that cholera was spread by a contaminated water supply. Quarantine the water supply, and London could limit the epidemic and save lives. Snow had hardly a clue as to the underlying cause, yet with this insight, the science of epidemiology was born.

The point is that both the idea of mapping and the linking of the mapping to the London water supply system (of all things) are acts of sheerest creation—surprising, unpredictable acts of grand synthesis—not analysis alone. And they are obvious only in hindsight.

So it is with great business strategies. We must remind ourselves, for example, that Fred Smith's hub-and-spoke overnight delivery concept, which gave birth to Federal Express, brought together a lot of new ideas. Smith saw that there was a widespread need for next-day delivery—a need most of us didn't know we had until he showed us that we did. He supported that idea with a dedicated airplane fleet, a center-of-continent nighttime hub, a company-owned fleet of trucks for major markets, and a bar code–based tracking system. It seems obvious now—we're all so familiar with it. But it's well known that when Smith promoted this idea in a business school paper, it garnered him a grade of C.[4]

Anatomy of a Great Creative Strategy

Ready to join in dissecting a great strategy? To understand the inherently creative core of strategy, let's dissect one of the boldest, most creative strategic retailing concepts of the last fifty years, the warehouse membership retailing concept pioneered by Costco and Price

Club. Most of us know this concept as customers. On the outside, it seems simple enough: Buy huge volumes and sell at low prices in a low-overhead, low-service environment. But the reality, as implemented by Costco, is far shrewder, more sophisticated, and much subtler. As a former vendor to Costco, I came to admire the sheer genius and subtlety of Costco's strategy. Figure 2-1 illustrates some of the disparate elements of Costco's strategic jigsaw puzzle that its founders, Jeffrey H. Brotman and James D. Sinegal, brought together to create enormous success:

1. Focus on small businesses for the main customer base—businesses too small to have corporate accounts with manufacturers and distributors but large enough to want to buy business supplies and related items in large quantities.

2. Serve the business owner, realizing that for small-business owners, personal and business needs merge. A trip to Costco for computer paper is a chance to stock the cupboard as well as the storage closet. So at Costco you'll find office supplies and office snacks but also fresh foods, wheelbarrows, books, and jeans.

3. Buy in large quantities to obtain favorable pricing but compensate for that by making the customer take large quantities, too. Often

Figure 2-1. Costco's complex, highly integrated, strategic if-then statement.

IF

a. We stock items for small-business owners' corporate and personal needs . . .

b. We offer many clues that the customer is getting a bargain . . .
- Name—Costco
- Warehouse atmosphere
- Membership exclusivity
- Payment by check or cash, no credit cards*
- Low unit prices on many items but large package quantities
- Minimal in-store customer service

c. We constantly change stocked items and their in-store locations . . .

THEN

a. We attract a larger share of small-business owners/ self-employed

b. We support an almost pure-profit membership revenue base at $40 per year per member**

c. We maintain over 75% membership renewal rate

d. We succeed in stimulating impulse buying—over 25% of purchases are unplanned (estimate)

* In 1999 Costco concluded special deals to accept Discover Card and American Express.

** "Executive" members pay substantially more.

this means requiring the vendor to provide special packaging: that extra-large jar of pickles, or a photographic film pack of a half dozen rolls of film, when the mass merchant standard might be only a four-pack. The customer gets a per-unit bargain, but the average ticket size goes way up. At the same time, require that the vendors package their items to fit standard pallet sizes (single or double pallets are okay, but only our pallet size, please).

4. Push the customer to make a Hobson's choice. Dispense with the traditional merchandising idea that a customer should have confidence in what your store carries. Instead, deliberately ensure that many items *aren't* available the next time the customer comes in. Over time, customers will begin to realize that they have to buy *now* to get the favorable pricing—there's no time to think or comparison shop. Such a policy allows Costco to take advantage of manufacturers' promotional deals. Since Costco doesn't feel compelled to keep items in stock, it can purchase from vendors on highly favorable terms, for instance, when there's a glut of garden hoses or garage organizing systems.

5. Offer almost no customer service to keep costs low and productivity high but also to reinforce the impression that members are getting a *good deal.* Costco's information systems know to the moment and to the penny how much of each item is in stock in which store, and how much of it was sold yesterday. But if you ask a Costco employee whether that outdoor grill you saw yesterday is still in stock, she won't know. The message: You should have bought it yesterday.

6. Provide highly favorable prices, especially on highly visible items, by keeping gross margins to the low double digits—something unheard of for most retailers. But more than make up for this by charging customers a membership fee for the privilege of buying at such low prices. Reportedly, two-thirds to three-quarters of Costco's operating profit is attributable to an item that has virtually no cost of goods—the membership fee. Costco's recent push to convert members to higher-priced executive memberships costing $100 and its raising of prices for regular and business memberships to $35 and $40 in 1999 are important parts of the strategy. Without the membership fee, Costco's sales would have to be on the order of 300 percent of current sales to generate the same operating profit.

7. In this era of credit cards and debit cards, Costco's cash- or check-only policy seems like a throwback to the 1960s. But it has important benefits. First, it dilutes the impact of those low gross margins by avoiding the 1.5 to 3 percent of sales fee that Costco's competitors have to pay. The financial importance of this policy is enormous: It increases Costco's margins by a typical 44 percent on an item being

sold for a target 15 percent gross margin. Small-business customers are accustomed to paying by check anyway, so little if any business is lost. Second, Costco collects its funds the day of the sale; there's no waiting for processing. And third, it increases store employee productivity. The checkout clerk never has to wait for the validation of Mr. Jones's credit card. (Note: In 1999, Costco forged special relationships with American Express and Discover Card to test the revenue and profitability impact of credit card acceptance.)

8. Finally, we come to the most visible element of the Costco strategy: those huge, bare-bones, unheated, poorly lit, warehouse-like outlets. They serve the strategy in several ways. First and most important, they send customers the clear message that they are getting wholesale pricing, because these big boxes look and feel like warehouses. Second, they are engineered for fast, efficient logistics—unloading, stacking, and displaying goods in warehouse racks is a snap compared with unloading and displaying in conventional retailing. Third, they're designed for overnight reconfiguration to allow space for new items. Often I've felt like one of those rats in a maze trying to find an item I *knew* Costco carried last week. Indeed, often a customer won't find the same item in the same place in two visits a week apart even if it's still in stock. This is a deliberate attempt to reinforce the customer's impulse to buy *now*, while the item can be found.

Costco's strategy is based on a perfectly falsifiable hypothesis: that there are enough small-business people who are willing to pay annual memberships, buy in quantity, and put up with a certain amount of inconvenience in return for very low prices (or the illusion thereof) to make this all work. Put in the target membership dues, state the typical prices, and specify the inconveniences, and you've got an exquisitely crafted, eminently falsifiable hypothesis worthy of comparison with Planck's constant hypothesis of chapter 1.

But for the purposes of this chapter, the Costco story illustrates how strategy is a sheer act of creativity. No strategic cookbook could ever have laid out the steps to this strategy. Of course, in retrospect, it seems obvious.

Guidelines for Creating a Strategic Hypothesis

The genius of hypothesis creation can be based on insightful observation, as well as on the melding of machinelike elements. The British scientist Alexander Fleming discovered penicillin, it is said, by accident. I prefer to say he discovered it by insight and imagination. The story goes that a dish in which he was culturing bacteria was accidentally left uncovered and became contaminated by a fungal spore. Fleming could easily have ignored the shrinking bacterial colony as a fluke. He could have noted it but assumed a simple explanation—an acid spill or contamination by something else found on the lab desktop—rather than hitting on the idea of invisible spores from the air (of all things). The hypothesis that some molds destroy bacteria was an act of sheer creation by a prepared mind. And it was testable: It could be made precise and falsifiable. No royal road to that, either!

Finally, the proof that you can't get strategy from this book or any other comes from the fact that the real world of business isn't written on the clean sheet of paper of geometry or played out in the controlled environments of physics and chemistry, where universal rules apply and remain applicable permanently. The real play of business depends to no small extent on the particular position your company has in its industry—big player or small, regional competitor or global, startup or tradition bound. The game of business is played in the forests, ravines, and mountaintops of production costs, sales force effectiveness, real estate costs, chief executive officer personalities, gross margins, advertising rates, customers' behavior, and government intervention. How could any book, software program, or framework take into account the infinite variety of the marketplace? Don't let anyone tell you that he knows one strategic premise is more important than any other, simply because he's created a successful business based (he thinks) on that premise.

For those of us whose creativity is less than cosmic and whose strategic issues focus on a business that exists in the here and now, despair is not in order. There is help. Just as our eighth-grade math teacher could provide us with approaches for solving those knotty algebraic equations, and our English teacher could suggest ways to plan and work toward the writing of a compelling essay, so too, history can provide us with guidelines for creating winning strategic hypotheses.

It's up to you to have the creative insight. You can't buy that—not anywhere. But I can show you how the skeleton of a successful strategy is constructed. And I can guarantee that if the strategic flesh you provide doesn't fit the skeleton, your proposed strategic hypothesis won't stand the test of the marketplace. If it does fit, you can be pretty well assured that you are on your way.

The First and Most Obvious

The first guideline is that a strategic hypothesis needs to be about something vital, crucial, and central to the company. Anything else is a distracting waste of time. This seems obvious.

But proposed strategies often fail this first important test. Judge for yourself.

Recently, Boeing's CEO announced that a review of the company's business found that 10 percent of its aircraft programs actually showed a negative return on investment. This review was the first big project of Boeing's brand-new chief financial officer (CFO), Debby Hopkins, hired away from General Motors. Said one leading aerospace analyst: "It's amazing that we are at this point and they are only now figuring out how to drop the losers and keep the winners."[5] Boeing's management had intense strategic discussions, I gather, but one has to wonder if they were about the most vital aspect of the business, since so many programs—programs for which Boeing's stockholders paid good money in the $13.4 billion Douglas deal—were money losers from day one.

Years ago, I attended a top-management presentation about the pricing and positioning of a trucking company's express services—strategic questions if there ever were any. The company, once a cash cow, was losing money hand over fist. It was being pressed between the millstones of low-cost competition from hundreds of Mom-and-Pop outfits on the bottom and the guaranteed, high-reliability services of huge, integrated truckers on the top. As a junior-level analyst, I sat in bewildered awe as the vice president of marketing presented the issue of whether the company's image was best represented by a bear or by some other animal. The hypothesis under discussion was certainly falsifiable. It just wasn't important at that time and place.

What *would* have been germane? Whether there existed a market segment of sufficient size and depth willing to pay midlevel prices for midlevel service, and just what those prices and service levels would have to be. And if that segment existed, whether our company could make money with that offering. And if it could, whether and how it could change its business system—the trucks, tracking system, sorting system, and productivity—to get from where it was to where it needed to be.

The test for meeting guideline number one is this question: If this strategy were implemented, would it be vital to the company? If the answer is yes, ask the sterner question: What would be *more vital* to the company?

Testing for Vitality

Once a strategy is proposed, ask the question What would be *more vital* to the company? three or four times, each time going to a deeper level of vitality. For Boeing, such an exercise might be as follows:

Initial Strategic Statement: If we eliminate low- and negative-return aircraft programs, we can return to respectable profitability. (I would call this a pseudo-strategy, but more on this later.)

Q: What would be *more vital* to the company?

A: Regaining our declining market share, not just profitability as a percentage of sales or return on equity.

Second Strategic Statement: If we eliminate low- and negative-margin aircraft programs, we can become a much lower-cost producer and regain the market share lost to Airbus, even though we would be ceding some markets to others.

Q: What would be *more vital* to the company?

A: Concentrating not so much on preserving older, profitable programs like the venerable 747 but developing competitive new products for the future.

Third Strategic Statement: If we reduce our costs by eliminating losing programs and update profitable products (e.g., the 747), the market will reward us more than if we launch entirely new products. (If more precisely defined, this becomes a falsifiable hypothesis.)

Q: What would be *more vital* to the company?

A: Not losing technological as well as market leadership in an era when each new product is so expensive that launching it is a *bet-your-company* proposition.

Fourth Strategic Statement: If we reduce our costs in the immediate future and use the funds to update current profitable products, we can stave off market-share decline long enough to forge financial and technological alliances with engine manufacturers, major airlines, governments, and avionics producers to launch world-beating products—at less financial risk than is involved in today's 767 and 777 programs.

It's an If-Then Kind of Thing—In the Market Place

The next most important guideline for a strategy is to realize that it is what logicians call a *conditional statement*—an *if-then* kind of thing. We touched on this in chapter 1, when we discussed how Kmart's and Wal-Mart's strategies are fruitfully thought of as falsifiable if-then statements. That's what a hypothesis is: *If* gravitational mass deflects light waves, *then* we should observe such-and-such displacement of Mercury's image when it passes on the other side of the sun from Earth.

The *if* part of the strategic hypothesis sets the conditions for the expected result. The *then* part of the strategic hypothesis sets out the results we expect—or hope for.

Obvious? Perhaps. But this requirement that we make the if-then connection explicit sheds a bright light on rules of thumb masquerading as strategies. For example, an oft-quoted *strategy* for General Electric (GE) is that the company wants to be number one or number two in any market it serves. I laud this as a *goal*. But is it a strategy? Perhaps it could be cast as a very high-level strategy if it were *unpacked* as follows:

> *If* we are number one or number two in every market where we compete (even if we have to buy our way into that position), *then* we are likely to earn higher than average returns to shareholders over a ten-year period.

This has the possibility of being falsifiable, at least if you define being number one or number two by market share (or operating profitability) and use industry historical financials to see whether companies in those positions earn those higher returns. Then you have the question of whether the high returns caused the high market share or the other way around. But let that pass for now.

Maybe this is what GE's chairman Jack Welch means by being number one or number two. Or maybe it's a rule of thumb used to signal managers: Don't bother me with any plans that don't meet extraordinarily high standards. Or maybe it's a rule designed as a hurdle for acquisitions or, conversely, a rule that identifies businesses ripe for divestment.

Most proposed corporate strategies flunk the *if-then* test. They're goals or missions, but they don't say why the company should be or will be successful if they're followed. Figure 2-2 presents some examples. On the left are some company goals and mission statements, like ones you've heard. On the right is what it might take to translate them into falsifiable strategic hypotheses—minus the *precision*, not to mention the *hard work*.

As you can see, the real strategy of these companies might well lie hidden behind the Potemkin village of the companies' noble missions. The strategy translations show how companies bet on the fulfillment of their missions—on satisfying what one philosopher calls the *covering* conditions of the strategy's *if* clause. And they bet that if they satisfy those conditions, they'll reap a reward; they'll gain a goal.

The chasm between the mission or goal of the company and its strategy also shows how important it is to clearly see and carefully consider a company's strategy. Look at the diabetes example. The orig-

Figure 2-2. Goals and mission statements are only starting points for strategic hypotheses.

Goal or Mission Statement	*Draft Strategic Statement*
A car for every purse and purpose	We will gain long-term market share if we provide auto buyers a ladder of models to purchase—from economical entry models to plush luxury cars—because people generally grow more affluent as they age
Not the biggest but the best networking software company in the world	We can charge premium prices and gain high market valuation if we provide a small portion of the market high-end networking solutions
The best deal in home financing for small-business people	If we develop unique expertise in the special mortgage loan needs of small-business people and keep hourly watch on the lowest loan rates offered by finance companies, we can cut prices and dominate this market segment
Helping people with diabetes live healthier lives	If we provide people who have diabetes with the ability to supplement daily, inconvenient, ambiguous blood sugar finger-prick tests with more convenient, patent-protected, less ambiguous once-weekly tests, health maintenance organizations will realize the health benefits and long-term savings and encourage their members to use our test regularly despite its relatively high cost (and high profitability to us)

inal premise of the company was that a test that showed people with diabetes how they were doing over a two- to three-week period would be highly valued as a supplement to today's glucose test. The latter test shows a person's momentary blood glucose level. But these levels fluctuate wildly from day to day—even from hour to hour—so it's difficult even for a diligent self-tester to know how well his or her regimen of diet, exercise, and medications are controlling diabetes.

You'd think that the new test would have blockbuster value for people with diabetes. Unfortunately, today the question in health care is less often What's the best way for me to treat my disease? and more often Who pays? An important part of the market—those patients served by health maintenance organizations (HMOs) and many preferred provider organizations (PPOs)—would be reluctant to enjoy that blockbuster value if they had to pay for the benefit out of their own pockets. Thus, an important part of the company's strategy becomes convincing HMOs that it's in their financial interest to promote the test among their membership—a far cry from "helping people with diabetes live healthier lives." And given the conservative nature of HMOs, the job of doing the convincing is a sobering challenge. Interestingly, for this company, there has to be another strategy—completely different—for the 20 percent or so of people with diabetes who aren't HMO or PPO members. Since the company is virtually in two businesses, it needs *two* strategies and at least two if-then statements.

If you insist on recasting your company's strategy as an *if-then* statement, you open the door to fruitful strategic dialogue. That's because the *if-then* statement breaks down the strategy into digestible pieces. We can stop arguing, for example, about whether a direct-sale-to-consumer strategy is better than a reseller-based strategy. Instead, decision makers can focus on the conditions (the *if* part of the proposed strategy). Facts can usually be gathered about the investment or change required to make the *if* portion true. Then we can discuss whether the *back end* of the hypothesis—the *then* statement, or consequences—is likely to follow from the conditions. And then we can discuss what everyone would accept as a *fair test* of the hypothesis.

This division of the strategic problem takes a lot of the heat out of those nearly theological strategic discussions so many of us have been involved in—discussions where everybody has an opinion and most of the opinions seem to have some merit, but authority and power tip the scales, and being provocative is a death sentence.

Even more important, the sharpening of alternative strategies into crisply defined *if-then* statements can boost the contrast between strategic alternatives. Here's an example: In 1996, Proton Computer's best sales, marketing, and engineering brains reluctantly concluded that they were chronically late to market with competitive products. The result was that they often had to slash prices to resellers in order to get them to take Proton's machines at all. At the same time, mass merchants such as Wal-Mart dangled the offer of purchasing many thousands of less-than-leading-edge products at very low costs. However, this entailed alienating Proton's computer reseller channel—already irritated at Proton's chronic tardiness to market.

A little background on Proton: A well-known computer brand, it not only sold through resellers but also had a loyal installed base and a name with a certain cachet. But in recent years the incomparably greater resources of Compaq (which also sells through resellers), direct merchants such as Dell and Gateway, and local no-name clones had squeezed Proton. At the time, Compaq, the world's largest personal computer (PC) maker, was nearly able to dictate the features and prices of PCs for the home and business marketplace by the sheer force of its market presence. Compaq could dictate, too, what the features of the next *hot* PC to hit the market would be. The second largest PC maker at the time, Packard Bell, was pioneering in the mass merchandiser market. Its strategy seemed to be to try to match the most visible features of the new Compaq machines but to provide a lower price by using lower technology and even less service.

By 1996, Proton's sales had risen beyond $2 billion, and its net losses had risen beyond belief. It faced a crisis as reseller margins shrank, product life cycles shrank, and its reseller market share shrank, while direct-sales companies grabbed more market share. Proton's near-majority shareholder, a Taiwanese consumer electronics maker, was getting very nervous.

What were Proton's strategic options? Its options could—and should—have been posed as a couple of opposing *if-then* statements:

- *If* we focus resources ($*x* million; *y* engineers) on new product development, *then* we will be able to nearly match Compaq in time to market with new products. This will allow us to beat Compaq on price by 3 to 5 percent, increase our market share 3–6 points, and eliminate our habitual losses on obsolete product liquidation. The results should be $*x* million cash flow, $*x* reduction in inventory, and welcome arms from our resellers.

- *If* we focus on supply-chain management (*not* new product development) and leverage our ownership relationship with our 49 percent Taiwanese investor, *then* we can offer world-beating prices for one-generation older computers to be sold at high volume and at thin margins. This will require that we shift our channels from high-service resellers to no-service discounters (such as Wal-Mart) and high-volume outlets (such as Computer City). But it will mean we can expect 25 percent greater volume at 15 percent less gross margin.

Proton's strategists could have developed either of these exciting hypotheses or others they hadn't conceived, like jumping on the personal digital assistant bandwagon. But because Proton's managers didn't see the two possibilities as sharply different choices about creating conditions (*if*) that would have probable and positive consequences (*then*), it tried to bet both ways. If managers had juxtaposed the two *if* statements, they might well have seen that expending time and money on cutting time from new product development would compromise investment in supply-chain management. Certainly they would have seen that an investment in squeezing time and cost out of the supply chain to support low-end products for resellers would jeopardize their channel. After all, Proton did not live in a world of unlimited time, talent, or money.

In the end, Proton lost both bets, along with its independence and $400 million in a single year.

The Pivot and the Hammer

So far we've talked about two reasonably obvious, if often overlooked, guidelines for creating strategy: making the strategy about something important, and making it a model of clarity by setting conditions and consequences—an *if-then* statement.

Now we come to a guideline that is not so obvious but is equally important, if not more so. I call it *seeking the Pivot and the Hammer.* As with other guidelines, the fact that your strategy *has* a Pivot and a Hammer is no guarantee that it is brilliant or even right. Furthermore, if your strategy *doesn't* have a Pivot and a Hammer, that doesn't mean it is doomed. But it's not a strategy I would bet on.

To understand the Pivot and the Hammer, let's take the first of our excursions into military history. In about 338 B.C., Philip of Macedon, Alexander the Great's father and unifier of the Macedonian tribes, began his campaign to conquer and unify the Greek city-states under Macedonian hegemony. The great British military historian and strategist J. F. C. Fuller paints Philip's discovery this way:

> He was a tactician of genius, and the first Greek general in history to grasp the importance of concentrating superiority of force against a selected point in the enemy's front. He realized that the Spartans were too conservative to change their traditional tactics, the success of which depended on an advance in parallel order, all spears of the phalanx strik-

> ing the enemy's front simultaneously, and he devised a system of tactics which would prevent this and throw the [Spartan] phalanx into confusion. It was the simplest of ideas; instead of drawing up his troops in line parallel with the Spartan army, he formed them into oblique order to it with his left leading and his right refused. And on his left wing he massed a deep column of troops that could meet shock by super-shock and possessed sufficient reserve force to lap round the enemy's right wing and drive it on to his center. In July, 371 BC, he used these tactics when he met and decisively defeated the Spartan army and killed their leader, King Cleombrotus, at Leuctra, in southern Boetia. This battle broke the charm of Spartan prestige, and ended Sparta's short-lived hegemony.[6]

It is a commonplace among military theorists that it is the soundest and safest policy, as Karl von Clausewitz says, "to be everywhere the stronger." But that is rarely, if ever, possible for armies or for companies. The next big thing is to have a concentration of force "at the decisive point."[7] A little logical reflection reveals that this *concentration* means de-emphasizing or, in plainer terms, *weakening* something else; only if you have comparatively overwhelming resources can you afford to be "everywhere the stronger." In business, the less charitable among us might consider that unpardonably wasteful.

The place where you decide to concentrate your effort I call *the Hammer*. The Hammer is where you play offense versus your competitors. The Hammer depends on *the Pivot*. The Pivot is where you play defense—where you hold the line in order to channel resources into the Hammer. In Philip's straightforward discovery, it's easy to see these elements. But almost everything is easy in retrospect.

On Philip's left, the Hammer: a concentration of a deep column of troops positioned close to the enemy, complete with a reserve force ready to go around the enemy to act as the Hammer.

In Philip's center, the Pivot: Philip's soldiers primarily in a defensive stance but linked to the Hammer. That is the reason for the oblique arrangement—to provide space for defense, ready to receive the enemy as they move naturally away from the momentum and power of Philip's left-wing Hammer, and to act in this case as an anvil for the Hammer. See Figure 2-3 for a schematic view of Philip's tactical innovation.

Here's another example from ancient Greece. About 140 years before Philip, the brilliant Athenian *strategos* (general) Themistocles provided a stellar illustration of this concept in one of the most decisive battles of history.[8]

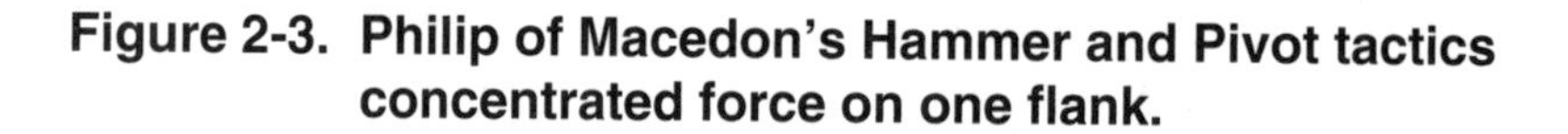

Figure 2-3. Philip of Macedon's Hammer and Pivot tactics concentrated force on one flank.

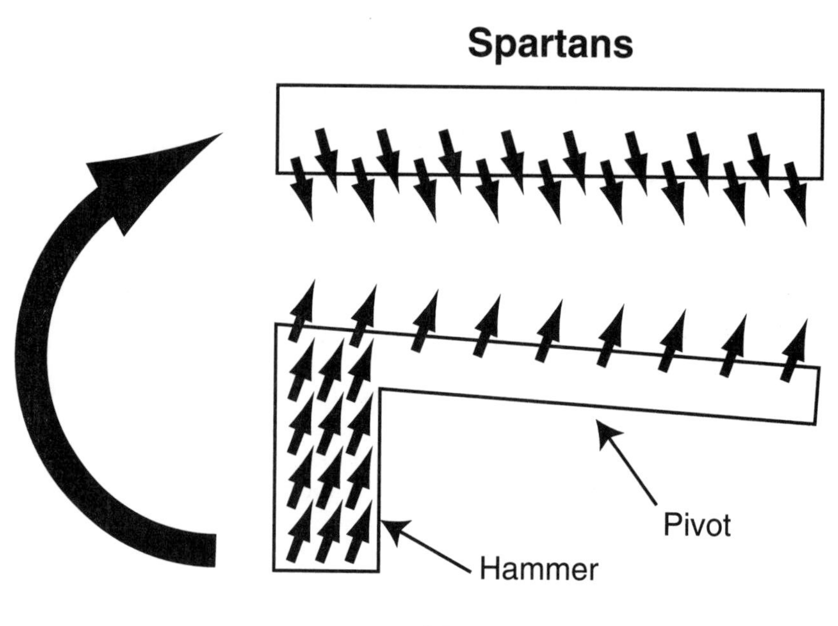

In 480 B.C., Themistocles faced a Persian invasion force of 160,000 troops—huge in those ancient times—that had annihilated the Spartans at Thermopylae, sacked Athens, and captured the Acropolis. It would be an understatement to call the Greek situation desperate. Except for a small Spartan army of 7,300, Themistocles had nothing left but an allied navy of 324 triremes and 9 penteconters. *Triremes* are fast, maneuverable ships powered by a total of 170 oarsmen in three rows, designed to ram and smash the opponent's oars and launch a boarding party of heavily armed marines. *Penteconters* are cargo boats with fifty oarsmen. The Persian naval force consisted of 1,200 warships.

First, Themistocles dulled the Persians' overwhelming numerical advantage by tricking them into attacking his navy in the Bay of Eleusis. The narrow Straits of Salamis would make the Persians attack in column, mitigating their numerical advantage. This trick allowed Themistocles to maximize his small force by narrowing the field of battle and concentrating what strength he had. That was the Hammer. But it was not enough. Themistocles had to create a Pivot—a way to increase the applied force of his Hammer—a fulcrum that played defense. He created his first Pivot by sending an allied, stripped-down

Corinthian detachment to the western end of the bay, where an Egyptian fleet threatened to attack him from the rear (see Figure 2-4).

Then Themistocles turned to the question of how to face the still-overwhelming primary attack by the Persians in the main channel. He decided to create a second Pivot for his Hammer. He concentrated most of his remaining forces on the Greek coast to his left, but he stationed a smaller holding force on the right—a force only barely capable of holding against the Persian onslaught. But hold it he did, while Themistocles' concentrated effort on the left mutated into offense. Thanks to this narrow concentration of his forces, the maneuverability of his ships, Greek seamanship, and the help of "Admiral Luck," Themistocles was able to work his way in among the Persian ships and rout them. The defeated Persians' sea communications with Asia Minor were so threatened that their king, Xerxes, pulled both troops and fleet out of Greece. The Golden Age of Athens was soon a reality, and the foundations of Western civilization emerged from it.

Why does the Pivot-and-Hammer strategy work? Let's take the Pivot first. The answer lies in understanding the inherent *power of the defense* on the battlefield. It's been known for thousands of years that it's generally easier to defend a position successfully than to attack one successfully. And the advantage is big. For years, the military rule of thumb has been that for a good chance of success, the attacking force

Figure 2-4. How Themistocles' Pivot and Hammer strategy overcame high numerical odds.

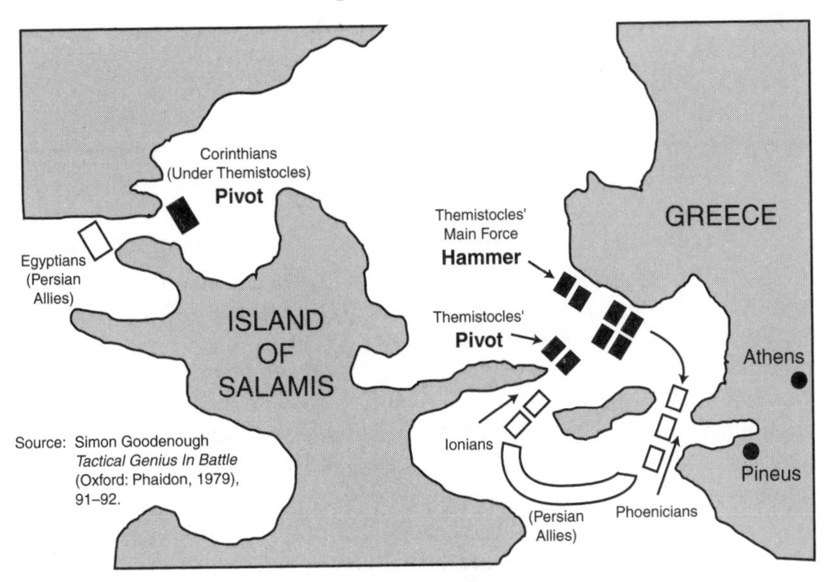

must have three times the power of the defending force. On difficult terrain, the rule of thumb changes to 6:1 or greater.

Some of the reasons for the power of defense are obvious; some are not. A defensive force usually has the advantage of time to prepare defenses—to build castles or fortifications or to set minefields (a modern favorite). It has time to learn how to take advantage of local geography to provide cover from attacking engines of war, be they battering rams or jet fighter-bombers. It has time to study the possible avenues of attack and prepare for those eventualities. It can build roads and communications infrastructures so that forces can be shifted from point to threatened point.

One of the more important but less obvious reasons for the power of defense is called "the advantage of interior lines of communication." As Figure 2-4 shows, a surrounded defending force can shift units and resources faster to threatened points on its defensive perimeter because it has less distance to travel.

According to a fascinating study, *How the North Won: A Military History of the Civil War,*[9] this fact alone accounts for the successes of the less numerous, poorly equipped, blockaded Southern states against the attempted invasions of Northern forces. Strategically, Southern states enjoyed *interior lines,* using their decrepit and scattered rail system to shift troops to the defense of threatened points. Tactically, embattled Southern forces assumed mostly defensive positions and enjoyed the same advantages. General Thomas "Stonewall" Jackson's classic Shenandoah Valley campaigns showed how an aggressive use of this advantage could give the appearance (in the days before ever-vigilant spy satellites) of being nearly everywhere at once and in overwhelming force at the chosen point. Yet Jackson's force was always outnumbered at least seven to three.[10]

"Only Make the Right Wing Strong"

Probably the most famous example of the Hammer and the Pivot—and the greatest cautionary lesson in why half measures are dangerous—comes to us from the strategy of the German General Staff at the beginning of World War I. Well before the outbreak of war in August 1914, the major European powers sensed that a fight must come in Europe, and they drew up strategies. The most interesting and instructive of these was that of Count Alfred von Schlieffen, head of the German General Staff and an archetypical Prussian, who died before the war's outbreak.

Schlieffen realized that although Germany was stronger than either France or Russia alone, once allied, the two countries would have

Germany virtually surrounded. If war came, he thought it imperative that Germany not fight a two-front war. Because of the vastness of the Russian nation, he thought it made sense to defeat France first, then shift resources to the Russian front. The victory over France could be achieved quickly (generals always think victory can be quick), and France could be forced to terms; then the German resources could deal with Russia before it had time to fully mobilize.

The result of years of monocled analysis and refinement was the Schlieffen plan. It envisioned defeating France by going on the defensive along the southern and central Franco-German border, especially in mountainous Alsace-Lorraine. In the north, however, Schlieffen contemplated concentrating the bulk of German forces for an overwhelming and swift march through conspicuously neutral Belgium, then hooking southward to capture Paris. Thus the right wing would be the Hammer that would pivot on the defensive left wing.

Schlieffen planned no half measures. To the defensive left and center, he sent 8 corps of 320,000 men and 11 corps of 400,000 men, respectively. But on the right, he concentrated 700,000 men in 11 corps. To do this, he planned something unheard of: calling up Germany's reserve forces immediately upon mobilization. There would be no reserves—all to make the right wing strong. Equally important, a relatively thin German force was to hold the frontier against any Russian forces until France was defeated in the west.

But on the eve of war, Schlieffen's successor, Count Helmuth von Moltke, simultaneously developed cold feet and overpowering greed. Moltke got cold feet about denuding the eastern (Russian) front so much. As for the forces in the south, the heart of the Pivot in Alsace-Lorraine, he began to hope that he could attack the French there, too. He knew the French were itching to recover these two provinces, which they'd forfeited to Germany in the Franco-Prussian war of 1870. The result was that he modified Schlieffen's plan, sending two corps to the Russian front and holding on to more troops on his left wing in Alsace-Lorraine to attack French fortresses. Schlieffen's 700,000-strong right wing dwindled to 500,000.

As every student of the war knows, the German right wing was stopped just outside Paris, along the Marne River, in the last days of August. The single largest and most desperate battle of the war was fought in its first month. In historian Barbara Tuchman's judgment:

> If the Germans had not withdrawn two corps to send against the Russians, one of the two would have been on [German general] Buelow's right and might have filled the gap between him and [German general] von Kluck; the other would have been with [German general] Hausen and might have

> provided the extra strength to overwhelm [the French forces under General Ferdinand] Foch.[11]

Further, Tuchman says, the attack on the French forces on the defensive in Alsace-Lorraine "meant committing the Sixth and Seventh Armies to frontal attack upon the French fortress line instead of keeping them available for reinforcement of the right wing. . . . Like the smile of a temptress, it overcame years of single, wedded devotion to the right wing."[12]

As an aside, the idea of the Hammer and the Pivot is still very much a part of the lexicon of military strategic thinking. In the Persian Gulf war, General Norman Schwarzkopf contained the Iraqi forces on the central front in Kuwait and used the U.S. Marines to threaten a seaborne landing on the Iraqi left. But the real power punch he reserved for a huge, sweeping left hook through the desert, descending in force on the Iraqi flank and rear. Only when that was accomplished did he permit an assault on the fortified Iraqi center, which had been pummeled into submission by weeks of air attacks.

Is all this talk of the power of defense and fortifications and the advantage of time and interior lines bunk when it comes to business? Quite the contrary. The analogies are remarkably apt, if not exact.

In nearly every kind of business where there's a chance of a repeat sale, it is far easier, less expensive, and more lucrative to keep an existing customer than to either convert a competitor's customer or acquire a new one. It's far less costly for Federal Express to keep me as a customer by performing well than for Airborne Express to convert me. Airborne would need to have a salesperson call on me (many times, before he or she caught me long enough to listen to the spiel), give me some free or discounted trial shipments, set me up with a new Web account and a new pricing schedule, and help me learn a whole new way of doing things for essentially the same end result (or value): a package delivered on time the next day, reliably, with tracking.

A business strategist's term for this is, as you know, *switching costs.* Switching costs are a powerful factor for the defense. Trout and Ries make the point in *Marketing Warfare,* without a discussion of causality. They emphasize how difficult it is for a new brand of consumer goods—say, beer—to overcome the position in the consumer's mind created by an established brand. The role of the challenger is to expend time, effort, and money redefining the established brand in terms favorable to the challenger (e.g., New Beer's the beer for the young and active; Old Beer, by implication, is for the old and inactive). Sometimes it works; sometimes it doesn't. It's *always* expensive.[13]

Beyond switching costs, a defensive position in a business market has a close analog to the power of interior lines. Suppose you are the regional sales manager for a large health plan. You're in Seattle and have ten account managers working daily with twenty large companies, helping them with their health insurance benefits. Headquarters wants to launch a new, costly initiative, involving the devil knows what in northern Michigan. Naturally, your boss in distant Dallas tells you that you've got to cut your budget to accommodate these grandiose plans. And on other expenditures, you're told to go over to an economizing *defensive* posture in your region.

You can do it. Although it's far from pleasant, you draw up plans for a 10 percent cut. Part one of your plan is sending Charlie, one of your less-than-stellar account managers, off to seek fame and fortune with the new initiative. Although you have 10 percent fewer managers, you've raised your team's average productivity by releasing this lower-performing resource (he needed a second chance in a new environment anyway, your transfer memo says). At a meeting of your nine survivors, you divide up the now-uncovered accounts. As a great sales manager, you've done those ride-along sales calls with Charlie and reviewed his weekly activity reports. So you know, to some extent, the key contacts in these accounts, their phone numbers, their specific needs, their likes and dislikes, and their ongoing issues. You can introduce Sue and Norman and Leslie to them, bringing them up to speed fairly quickly. Since you're in defense mode, you cut the amount of time Sue, Norman, and Leslie spend on cultivating new accounts and maybe drop their frequency of visiting existing accounts by 5 percent.

Result: a smooth hand-off, fewer customer complaints, and no accounts lost.

You knew the territory, your communications system was intact, and you could transfer resources to cover what had to be covered. Sure, everybody had to work a little harder, but since you didn't have to start from scratch to save the $200,000 blood offering headquarters demanded, the job was doable.

Compare this with the extra time, effort, expense, and risk you'd have to expend to gain the same dollar amount of value (that $200,000) from selling, converting, setting up, and learning new accounts. To gain that kind of incremental gross margin in many industries, you'd have to find *new* customers worth $1 million in annual revenue. You might have to invest $500,000 in travel, sales time, and your time to capture those new accounts.

The difference between that $500,000 and what you spent in time and effort to have your surviving account reps cover Charlie's accounts is the *economic value of defense.* It's worth calculating. I've summarized the power of the defense—why the Pivot can work—in Figure 2-5.

Figure 2-5. The power of defense in military strategy and in business.

Feature	*In Military Strategy*	*In Business**
The Advantage of Time	Defender knows the geography, has time to build fortifications and communications, and can study likely avenues of attack	Defender has been in place; knows customers' needs (or knows how to manufacture at low cost)
Economy of Resources	Defender can use fortifications to multiply inherent firepower	Defender can leverage through use of existing customer relationships and knowledge of customers' specific needs to avoid wasting sales and marketing (or knows how far manufacturing process can be economically tailored to meet sales needs)
Established Infrastructure	Defender can lay roads, communications network, intelligence network to speed the transfer of information and firepower to the place where they are most needed	Defender has an established communications routine with customers, methods of finding out customer needs, and might even know by heart the airline schedule between major cities (or, in manufacturing, has long-proved knowledge of best practices, how to get best prices and service from parts vendors, has a proven and refined fabrication process layout)
Knowing the Territory	Defender knows the roads, rivers, forests, and mountain passes that an enemy could use to approach the fortress	Defender knows the ups and downs in the local market, the location of competitors' offices and salespeople, and the weaknesses and strengths of their product offerings (or, in manufacturing, knows which vendors can be relied on, where to find skilled labor, how much to pay, and where to find repair services)
Interior Lines	Defender is located centrally with respect to the perimeter being defended. He can quickly switch forces to wherever there is a threat using a dense, short-distance road network	Defender is located in the heart of the market and can easily reach and reassure key customers, switching sales efforts to where they are needed, whereas attacker must spend energy and time locating, contacting, and getting appointments with numerous potential customers before finding one who's interested (or, in manufacturing, knows already how to allocate resources to different product mixes as customer needs change)

*The advantages of defense are illustrated assuming that the marketing function is chosen to go on defense. But in reality, any company function can adopt a defensive stance (e.g., manufacturing, research and development, sales, customer service, finance).

But, you say, you can't grow a business by shrouding it everywhere in the cloak of defense. You are absolutely right. You might not even be able to survive long term. As competitors gain share while you're on defense, they might well experience increasing dollar profitability. They can reinvest some of that money in putting sales feet on the street in your Seattle territory. They can launch their own initiatives in northern Michigan, longer term.

Indeed, there are few situations in which it pays to adopt a long-term, businesswide defensive posture. This might be called "all Pivot; no Hammer." When companies do it, it's called *harvesting,* and it expresses the notion that it's better to return profits to shareholders than to invest or reinvest in the business. But the defense is often so strong that companies on defense can linger for a long time. Woolworth's is an example of a company that lingered and shrank for years before closing down altogether.

What is the Hammer? It's the central force of your main effort. It's where you create or exploit an advantage. It's what you use to overcome competition, capture customers, and build markets. The Hammer is the engine of growth. And it is the embodiment of the strategic hypothesis on which your company's hopes, dreams, and plans depend.

Richard Foster makes the case that every company should have an offensive posture in today's world in *Innovation: The Attacker's Advantage.*[14] Foster makes two main points: First, innovators, especially technology-driven innovators, don't have the commitment—technical, human, and emotional—to an older way of doing business. The past is no millstone around the innovator's neck.

Second, marginal investment in a new, valuable technology will create higher value for customers (and therefore profits or market share) than the same investment in an older technology. That's the power of the famous S curve for the adoption of new technologies. So, for example, when steam power for ships began to challenge sail, the response of sailing ship builders was to add more and more masts and more and more canvas. But for all that effort, sailing technology was exhausted. The law of diminishing returns had set in: Added masts and added canvas added little speed. Despite their beauty, majesty, and grace, the multimasted clipper ships of the 1870s and 1880s were only marginally faster than their predecessors, and by the early 1900s, they were fully overcome by steam.[15]

Foster's focus is technology, but the same argument can be made without appealing to the peculiarities of technology-driven industries. If you have a good, long-standing customer in Sally, and you already have a good portion of her business, it's likely that you'd have to devote half again as much sales effort to get another 10 percent out of her. It might be better to redirect your sales resources—salespeople's time, sales support, promotional materials—to more deeply penetrate new accounts, where the same effort might provide larger gains. Poor Sally: She'll have to be content with the same number of free theater tickets she got last year.

Retailing shows the same S pattern. The hardware store buying groups such as Ace Hardware, Coast-to-Coast, and Tru-Value are scared to death that what they call the *"big boxes"* category killers such as Home Depot, Loewe's, and Home Base will make their members and customers—local hardware stores—a thing of the past. And they're right. For better or worse, and sadly enough, incremental investment in the old way of doing things will yield lower returns than investing in Home Depot stock. The next ten years will show whether these groups can transform themselves radically, or whether they will merely linger, the past a millstone around their necks.

Another example: The Internet is changing the way music and mortgages are bought. In the music world, unknown artists can post their recordings on sites such as *www.mp3.com.* Visitors can listen to selected tracks free of charge, then download what they like or order customized CDs that contain only the tracks they like from the artists they like. It's a tiny business. But if you were a top executive at a big record label company, would your comment on this phenomenon be, "Blah-blah-blah-blah-Net, blah-blah-blah dot.com," like that of Val Azzoli, one of the cochairmen and co-CEOs of star-maker Atlantic Records?[16]

If you were to survey people about desirable ways to spend time, shopping for a mortgage and getting the straight scoop on rates, points, appraisals, fees, and maximum monthly payments probably ranks right up there with buying tires and scheduling root-canal work. But unlike those actions, the process of shopping for, applying for, getting approval of, and closing on a mortgage is almost purely a transfer of information—information about salaries, assets, liabilities, home values, interest rates, and various fees. What a natural for the Internet! How long will it be before the majority of mortgages are originated and perhaps consummated electronically? How many storefront mortgage brokers will there be in 2005?

Offense is usually necessary for survival, but it's an absolute requirement for growth, and it's more fun. But as we've seen, it's also more expensive. So the Hammer's success depends not only on its own cleverness, energy, and timing but also on the cleverness, parsimony, and commitment of the people at the Pivot.

Companies use the Pivot and the Hammer all the time. That doesn't make it right, but taking the foregoing logic into account, we can see why it works. For example, in the 1980s and 1990s, Microsoft used its market power in operating systems as a Pivot point on which to Hammer one software sector after another—Lotus in spreadsheets and integrated office suites, WordPerfect and others in word processing, Novell and others in networking. It hammered and hammered away until it fully dominated the office suite software market and had knocked Novell from its pedestal as unchallenged network leader. I'm told that in 1998, Microsoft Office made up fully 40 percent of Microsoft's revenues. Talk about Philip of Macedon's concentrated resources!

When Microsoft ultimately decided that the Internet was for real, its programmers did an about-face to create first a clunky browser, then a more competitive one at very high cost, and then distributed it free of charge. The company is no doubt pivoting on its Office revenues. Resources that would have been spent improving Office are now shunted to Internet, multimedia, and e-commerce ventures.

Coca-Cola pivots on its flagship soft drink products while moving into the bottled water market, launching the Dasani product. The Pivot here is not just Coke Classic's cash but also the company's clout with grocery store chains. Coca-Cola has no problem at all getting prime, eye-level shelf space for its version of filtered water.

Chrysler (before the Daimler merger) was number three and played defense in the sedan market while focusing energy, attention, and innovation on the light truck and sport-utility market. Chrysler is, they say, "a truck company that makes a few cars." It probably loses money on cars. Why does it keep making and updating them? They're the Pivot on which the truck strategy rests. Dealers need them. Future car customers would be wary if Chrysler abandoned such a visible market. (And then there's the nagging possibility that one day, the SUV and truck boom—fad?—will fade away.)

It may not be that every sound strategic hypothesis has an identifiable Pivot and Hammer. But it's the way to bet.

The Principle of Complementarity

It's time for a rest from war and business. Let's take a momentary excursion to your local art gallery. Your taste runs to the modern, the clashing, the abstract, the exciting with a twist. You turn away from the cliché. But today, you're looking for something that Aunt Mildred might like. You pause for a minute to contemplate a painting you catch yourself *enjoying* ever so slightly, and you are a little annoyed with yourself. The painting is *representational*: It depicts something more or less recognizable; it isn't abstract. Worse, it is a cliché, painted summer in and summer out by amateurs and professionals alike. A red barn with rooster weather vane graces a green summertime New England field. While glancing furtively this way and that for the art police, you almost admit to yourself that you *like* the painting. You remark to yourself, "How lovely for Auntie's dayroom!"

A stone wall—stones so perfectly depicted you can feel the strain of the farmer's muscles lifting them into place—starts at the picture's lowest edge and carries the eye to the barn. A survey of the sky reveals that the clouds, too, are carefully placed to frame the barn. The painting's reds, greens, and blues are related in an easy, pleasing harmony. You can see the details in the grasses and flower blossoms *close* to you; then all those details blur and fade to provide the illusion of depth. The Berkshire Hills form a distant, misty, vaguely purple backdrop to the middle-distance barn and fields.

There is a good reason you enjoy the painting, despite yourself. Each of the elements complements all the others. The painting's colors are a classic triad known to work well together. The placement of clouds and stone wall leads the eye to the focal point, the weathered barn. The object of maximum interest is also the locus of maximum contrast of light and dark. It's also the locus of maximum color contrast. All this leads to the enhancement of the painting's focal point, the red barn. The shapes of the main elements are all designed to lead the eye toward the barn, and equally designed not to steal the show from it.

A sound business strategy is typically as complementary in its elements as the elements of that painting. This guideline is perhaps best explained by illustration. In fact, let's use the Costco example again, a strategic work of art if ever there was one.

- *Selling* to consumers—sorry, at Costco you have to call them *members*—in larger-than-normal quantities at low per-unit (but high per-ticket) charges complements *buying* in larger-than-normal quantities (at low per-unit costs).

- Store—sorry, at Costco you have to call it a *warehouse*—size, shape, and rack design facilitate fast unloading of products from vendors. This complements the presentation of an ever-changing experience to members and supports the underlying message that *they'll never pay retail* at a Costco warehouse.

- The membership fee and the cash- or check-only policy, which weeds out poor credit risks, complement the marketing strategy by targeting small-business people. The cash and check policy—no purchase orders—means that Costco often (always?) gets paid for its goods by customers before it needs to pay its vendors, which is great for cash flow.

- *Requiring* vendors to provide Costco with *unique* SKUs that fit its standardized warehouse pallet dimensions complements Costco's marketing message that you can find items at Costco that you can't find anywhere else. And it complements the strategy of *redesigning* the interior space frequently and quickly.

The original Ford Taurus is also a case for complementarity in a strategy. Yes, it was researched and designed to meet the Middle American market. But it was also designed and priced to leverage Ford's relations with fleet buyers—large companies and rental car companies—that had their own requirements. In later models, the modularity of its interiors (notably the ovoid dashboard) was a boost to manufacturability, but it also provided a *family resemblance* among all Ford's passenger car products—a look that marketers cherish.

A Checklist: Putting the Guidelines to Work

To craft a successful strategic hypothesis, begin by asking yourself the questions in Figure 2-6. Just as following the guidelines for a good painting, short story, sculpture, or three-act play won't guarantee the creation of a crowd pleaser, following these guidelines won't *guarantee* a sound strategy. And you'll find that some strategies work even

Figure 2-6. Asking the vital questions about your company's strategy.

Question	*Your Answer*
1. Is today's strategic hypothesis about something vital? Does it go to the heart of the company's reason for being? How?	
2. How would you state the company's strategy as a crisp *if-then* statement?	
3. Does it have a recognizable Pivot? A recognizable Hammer? Where are we piling it on? Where are we playing defense?	
4. Are the elements of strategy self-reinforcing—complementary? If so, how?	

though they flout one, two, three, or all four of these guidelines. But it isn't the way to bet.

You may be saying to yourself right now, My strategic hypothesis is clear; the conditions and consequences are spelled out, falsifiable, and testable; there's a Pivot and a Hammer, and I've stripped where I can to reinforce the Hammer; and all the elements support one another with a minimum of conflict.

But just what does the Pivot look like? Where—now—do you direct the Hammer to fall? Just how do your Pivot and your Hammer work together?

Notes

1. Barbara Tuchman, *The Guns of August* (New York: Ballantine Books, 1964), 42.
2. Michael Cowley and Ellen Domb, *Beyond Strategic Vision: Effective Corporate Action with Hoshin Planning* (Boston: Butterworth-Heinemann, 1997), 61–62.
3. W. A. Wickelgram, *Problem Solving Strategies in Mathematics* (1974; reprint, New York: Dover, 1995).
4. Vance H. Trimble, *Overnight Success* (New York: Crown Publishers, 1993), 80.
5. Laurence Zuckerman, "Boeing Weighs Tough Steps to Increase Profits," *New York Times*, 25 February 1999.
6. J. F. C. Fuller, *The Generalship of Alexander the Great* (1960; reprint, New York: Da Capo Press, Plenum Publishing, 1989), 22.
7. Anatol Rapoport, ed., *Carl von Clausewitz on War* (New York: Penguin Group, 1988), 266–68.
8. Simon Goodenough, *Tactical Genius in Battle* (Oxford: Phaidon, 1979), 89–92.
9. Herman Hathaway and Archer Jones, *How the North Won: A Military History of the Civil War* (Chicago: University of Illinois Press, 1991).
10. Ibid., 176.
11. Tuchman, *Guns of August*, 485.
12. Ibid., 265.
13. Al Ries and Jack Trout, *Marketing Warfare* (New York: New American Library, 1986), 137–54.
14. Richard Foster, *Innovation: The Attacker's Advantage* (New York: Summit Books, 1986).
15. Ibid., 27–28.
16. Jodi Mardesich, "Features/Music & the Net: How the Internet Hits Big Music," *Fortune*, 10 May 1999, 96.

Chapter Three

Strategic Anatomy: Strategy's Hammer and Pivot

My formula for success? Rise early, work late, and strike oil.
—J. PAUL GETTY

A desk is a dangerous place from which to watch the world.
—JOHN LE CARRÉ

Of the four strategy guidelines discussed in chapter 2, one stands head and shoulders above the others in importance: the idea of the Hammer and the Pivot. To understand how to fashion a robust Pivot and a powerful Hammer, it is useful to dissect their anatomy. It just won't do to simply proclaim, as many chief executives have, that "this year, marketing (or sales, or engineering) will be our focus, while we continue our proud tradition of superior customer service." Everyone knows that this kind of phrase is code for belt-tightening, cost reduction, and job cuts in those areas of "proud tradition," while the other functions get the glory.

Instead, we must understand the particular nature of and requirements for a functional area or business unit that can act as a successful Pivot or Hammer to a strategy. This is akin to those college biology labs, where, once you got past the smell of formaldehyde, you could see the marvelous intricacy of ball-and-socket joints, the infinitely complex yet perfectly efficient network of blood vessels, and the complementary muscle systems of the dissected specimen.

In this chapter, we'll explore the anatomy of the Pivot, the anatomy of the Hammer, and the main ways they work together to fashion a competitive strategy.

The Anatomy of the Pivot

What is a Pivot? It's those aspects of our current business that *we rely on* to *hold their own* while we shift resources to create the conditions called for by our strategic hypothesis. That much is intuitively clear from our discussion in the last chapter. But hold on—let's unpack this innocent-sounding definition.

"Aspects of the current business" can mean functional areas such as finance, control, manufacturing, or sales. It can mean main business units that are steady moneymakers. Or it can mean a brand name with a prominent place in customers' minds. Or geography-based business units.

"Rely on" requires a bit more unpacking. First, it means that these particular aspects aren't subject to stresses and strains of the same magnitude as places of strategic interest—for the period of the strategic plan. For example, saying that we can "rely on" finance means that finance will be able to do its job pretty much on the trajectory it's now on for the next five years. This is far from a no-brainer: Many times, finance is actually an important part of the Hammer. An example is when a company exploits its high stock price to make crucial acquisitions or reengineers its balance sheet.

It's not just functional areas that can be relied on. Steadily productive business units can also be parts of a Pivot. They're the famous *cash cows* used to fund more exciting and lucrative opportunities elsewhere. Other assets can be relied on too. For example, many a brand name has become the Pivot of a new strategy—for good or ill—such as when Coca-Cola extended its brand to New Coke, Diet Coke, and Cherry Coke.[1] Another sort of asset that can be relied on may be a company's position with a channel of distribution or its relationship with key customers. For example, Atlantic Records won't have any trouble getting shelf space for CDs in music stores for a major artist it's trying to push. Once the ads, interviews, and other hype tools are in place, space will be found.

Now what's important is that for some function, business unit, brand name, or other asset is to work as a Pivot, the new strategy must not (significantly) degrade that asset's reliability. For example, if a new strategy relies on the charmed position a brand name holds in the minds of consumers, new products launched under that brand name had better not cheapen, dilute, or confuse the current brand name. If a strategy relies on the cash from a steady business unit, siphoning of that cash had better not lead to (significant) degradation of Steady Eddy, at least until the new strategy takes hold.

The Pivot is not simply everything we *take for granted* while resources are lavished elsewhere. Bookkeeping is typically something we rely on, but it is rarely a Pivot. The functions, resources, and assets of the Pivot are actively managed to protect them while they provide resources to the Hammer. In chapter 2, we saw how Themistocles thinned his naval forces on one side of the channel where the Persians were attacking so that he could concentrate his forces on the other. That didn't mean that the forces he pivoted on did nothing—far from it. They fought desperately on defense and attempted to engage as many of the Persian fleet as possible. Similarly, Schlieffen's left wing along the Alsace-Lorraine frontier wasn't intended to merely sit there. It was to be an active threat to the French forces lying opposite.

Likewise, a technology company with a *fast-follower* stance uses engineering and product design as a Pivot, but a very active one. For example, Matsushita's Panasonic brand has traditionally held to a fast-follower competitive stance. It doesn't need leading-edge product innovations—it lets Sony and Sharp do that. As the market for the latest electronic gizmo becomes real, Matsushita copies, and sometimes improves on, and often reduces the price for comparable equipment. Matsushita pivots on its engineering and manufacturing teams.

A related example is Southwest Airlines' continuing expansion of its network. Southwest built its business on low-cost service. This it achieved through nonunion, low-cost wages with employee stock ownership; acquisition of new, low-maintenance equipment; and standardization on just one aircraft type. Southwest's Pivot has two elements: its low cost position and its current portfolio routes. It takes active hard work to maintain those, and as Southwest expands nationally into routes for which its equipment is perfectly suited, it may be risking a portion of that Pivot.

The Bearing Inside the Pivot

Our dissection of the Pivot reveals a crucial element: a Bearing. The Bearing carries its burden smoothly, distributes the stresses, and eliminates destructive friction. In a jeweled watch, for example, the jewels are the highly polished, almost frictionless Bearings on which the moving parts rotate. In a playground seesaw, a large metal Bearing carries the entire load of the balance when both children are airborne.

Not every Pivot in a Hammer-and-Pivot business strategy has a Bearing that is the crucial, indispensable supporter of the load. But it wouldn't be wise to bet against the existence of *at least one* Bearing in any good Pivot. The more deeply you probe into a successful company's Pivot, the more likely you are to find one or two—sometimes tiny—assets, skills, people, or cultural factors that the company relies on to carry the entire defensive burden. This is true even in huge companies.[2]

A fairly large market research firm I know pivots on the specialized quantitative knowledge of just a single member of its professional team. Nearly all the firm's assignments rely on one powerful statistical methodology. Some very large companies, such as the bigger automotive companies, use it. At first, we might think that this methodology is the Pivot's Bearing. However, if we go deeper, we find that it is the single professional's ability to translate the results of the technique from academic research into live business applications that is the real Bearing. Dissect a little further, and we find that this professional recently created a unique template for a computer model of the technique that empowers the firm's clients to manipulate, test, and otherwise *use* the answers provided by this highly complex technique.

I've been in meetings with this company's clients. What makes their eyes pop out is the handover of this statistically complex model in a simple-to-use personal computer format. The model empowers them to play what-if games to their hearts' content, testing and modifying business plans and projections in myriad ways, no longer relying on reputations, elegant arguments, or avalanches of *confirming evidence* to guide them in the preparation of marketing strategy.

The real Bearing in this company's Pivot is the professional's creation of interactive PC models for the easy but highly productive use of the market research company's statistical findings. The original program probably took no more than a few research brain-months to create. But like the jewels in the watch movement, these programs are immensely important to the success of the firm.

Let's take another kind of example. A few years ago, a large, highly successful department store retailer was making forays into new markets, leveraging the steady cash flow of its existing stores. Those stores were (and still are) the Pivot of the company's strategy, providing the cash, the cachet, and the buying power to support new strategic efforts.

But if we brandish our dissecting knives, the conundrum immediately suggests itself: What makes this Pivot work? Granted, these stores have been successful in the past. But what makes us think that we can depend on their continued success—at least until the new strategies take hold?

Many of the stores have pretty good locations. They are fairly good sized, so they get reasonably good pricing when they meet the vendors across the negotiating table. But my investigation suggests that all this would rapidly go for naught if it weren't for the troglodytes inhabiting a couple of dozen windowless, cheaply paneled offices in a large city in Florida. This dedicated corps of highly competitive, frighteningly savvy, maddeningly confrontational retail buyers stares for hours every day into antiquated computer monitors, using obsolete software. They talk for hours on the phone with vendors and manufacturers and deal daily with the company's spaghetti-like network of distribution centers. Yet they succeed every day in arranging to offer exactly the right mix of merchandise to an ever-changing mix of ethnicity, age, and other demographic ingredients.

This corps is managed, motivated, and mentored by an energetic, balding veteran of several regional department store chains. A thorough dissection of the Pivot might find that his knowledge and personality constitute this company's crucial Bearing. Even if the buyer corps were destroyed through some commercial holocaust, this manager has the uncanny knack for recruiting people from campuses and malls that he could mold into his own team of spectacular retail buyers.

These examples provide the most important criterion for identifying the Bearing inside a strategy's Pivot: The Bearing is that which is absolutely necessary, perfectly essential, and practically sufficient to constitute the entire Pivot of a company's strategy. It is the essential asset that, if it were destroyed or missing, couldn't be reconstituted quickly enough to save the Pivot.

These examples also show that it's almost always possible to dissect more deeply, understand more profoundly, and narrow down more precisely the Bearing inside the Pivot. I challenge you to push this analytic thinking as far as you can. It will be far from an idle exercise. Seeking the Bearing in the Pivot allows you to

- Identify precisely what assets your whole strategy depends on. In the last example, a tiny group of perhaps fifty souls in a corporation of 12,000—and within that group a single individual—holds the keys to the company's future.

- Create contingency plans so that your strategy is never held hostage by the crucial Bearing. The department store retailer's strategy is hostage, to some extent, to its individual Bearing's health. And if ever there was heart attack material, this man is it.

- Decide where you should concentrate the resources you've allocated to defense. In our example, making sure that the buying group was robust—and perhaps supplied with updated information technology—might be a wise, if unglamorous, investment. The Hammer in this company's strategy was targeted to strike in exotic venues such as catalog sales and overseas investment. Without careful Pivot analysis, how easy it might be to deprive the tiny buying group of even modest information technology upgrades that could help ensure that they would continue to deliver shoppers, revenues, and profits to support the new strategic thrust.

Some might argue that there is no reason why there should be one and only one Pivot to a strategy. And they would be absolutely right. In the Costco example dissected in the preceding chapter, a close investigation might find two or even three. Currently, Costco's strategic hypothesis appears to be that if Costco can properly identify its most affluent members, then through the use of high-priced executive memberships, it can dramatically raise revenues per store—selling not higher-ticket merchandise but high-value services, such as automobile brokerage, real estate brokerage, insurance, and mortgage loans.

This strategic thrust depends on maintaining and even expanding the core high-volume merchandise sales. That business is Costco's Pivot for this strategy. The Bearings for this Pivot are the vendor relationships (high-quantity, low-unit-price SKUs), unique store logistics (store layout and warehouse rack design), and credit policies (membership dues and payment by cash or check only). I'm sure that if we were to dissect each of these Bearings, we could fingerprint them even more precisely. For example, if we looked into the logistics Bearing, we would probably find a unique-to-Costco rack design that allows quick disassembly, transport, and setup. At a minimum, Costco has strict specifications—to the inch—for the pallet sizes vendors are allowed to use for shipping goods to Costco.

So far I've used mechanics—the science of simple machines—as an analogy to bring to mind the concept of the Bearing. And we have inspected representative types of Bearings in several business situations. But a broader perspective reveals that even nations often rely on such Bearings for their strategic success.

In his mammoth book *The Wealth and Poverty of Nations: Why Some Are So Rich and Some So Poor,* Harvard professor David S. Landes shows how England's security and economic well-being in the seventeenth and eighteenth centuries relied on its command of the sea.[3] We're talking here not only about the ability of the Royal Navy to project force to the European continent but also about its ability to conduct worldwide commerce in fierce competition with European powers. That ability relied on superior navigation skills (and, to be sure, other skills, cultural factors, and England's implementation of the rule of law).

But England's competitiveness—even superiority—in navigation relied in no small part on the advanced talents of its watchmakers. Precision in chronometrics was the crucial ingredient in navigation, and England made the best clocks and watches in the world.

In her little book *Longitude,* Dava Sobel shows how for almost a hundred years the English crown pressured scientists and craftsmen to solve the problem of determining a ship's longitude. Ships, men, commerce, and time were being lost to navigational reefs and shoals, Spanish men-of-war, and other avoidable hazards of the sea. In the end, the book's hero, watchmaker John Harrison, solved the problem by crafting exquisite chronometers using what might best be described as *alloys* of woods in the key parts. The composite nature of these wooden parts compensated for the varying temperature and humidity conditions ships experience, allowing unparalleled accuracy in time-keeping. Equipped with these tiny Bearings, the British fleet and the merchant marine enjoyed an important, decades-long advantage on the high seas, during the explosive years of the industrial revolution and European empire building.[4]

Another example of the existence and identity of Bearings within Pivots in history is provided by Stephen Ambrose's book on the Normandy invasion of World War II, *Pegasus Bridge*.[5] Everyone knows the story of the storming of the Norman beaches code-named Utah, Sword, Juno, and Omaha. Those beaches were chosen in large part because the bulk of the defending German force wasn't defending them. The biggest, most aggressive, and most heavily equipped German divisions were located farther north, on the invasion's left flank. The Allied invasion plan had to forestall any possibility of a crushing German counterattack on that left flank. The invasion's most vulnerable hours would be those of the first day, when Normandy's beaches would be crowded with tens of thousands of men, vehicles, and artillery pieces milling about as they grew strong enough to crack the beachfront fortifications and break out from the beaches.

The Pivot for the entire invasion strategy was to plant a defensive force by parachute and glider along the Orne River and the River Dives. These rivers formed the only real natural barriers to a German counterattack from the north. However, glider and parachute troops themselves are by nature few, lightly armed, and disorganized when they land. Until they get organized and get to their supplies and heavier equipment, they are no match for a quick, well-led counterattack by well-armed defenders.

The security of the flank fell under British responsibility. The Bearing to this Pivot turned out to be a single bridge across the Orne River—the fastest, most direct pathway for a German counterattack on the Normandy beaches. The British created a special six-platoon company under Major John Howard, whose mission was to seize the bridge intact if possible, destroy it if necessary, but at all costs defend the area in the hours between midnight and dawn on 6 June 1944, until the Pivot could get itself organized. According to Ambrose, it is fair to say that on this tiny group hung the entire fate of the invasion force (see Figure 3-1).

And the fate and success of this tiny band, it turned out, rested largely in the hands of a single man, Sergeant M. C. Thornton. British intelligence knew that a German unit, including tanks, was billeted in

Figure 3-1. Normandy invasion pivoted on a single crucial bridge.

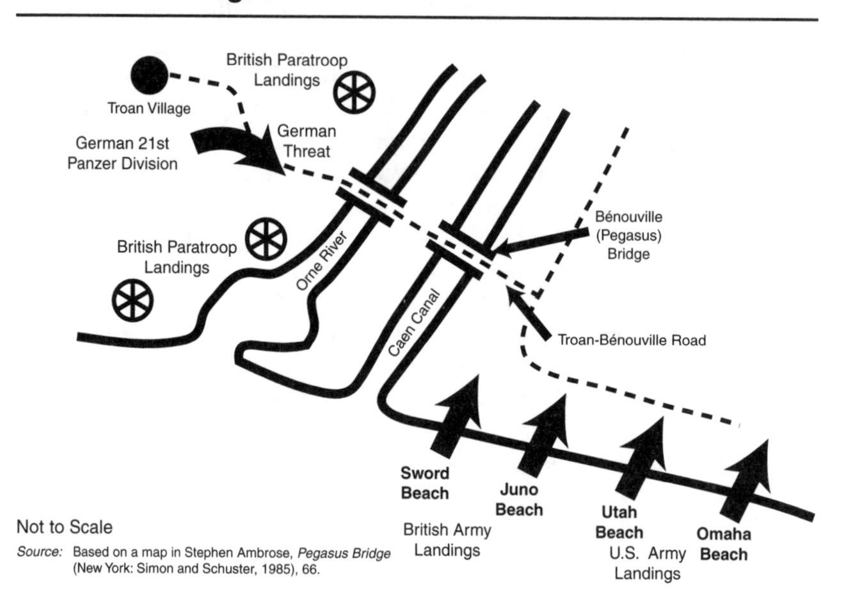

Source: Based on a map in Stephen Ambrose, *Pegasus Bridge* (New York: Simon and Schuster, 1985), 66.

the tiny town of Bénouville, a few kilometers from Pegasus Bridge. As soon as word of the coup-de-main seizing of the bridge reached this unit, Thornton's group could expect an immediate attack with far more firepower than the airborne defense unit could ever expect to muster. The best hope they had for survival was some appallingly short-range, shoulder-carried antitank rocket launchers called Piats. Grenades fired from a Piat were so short ranged and the Piat was so slow to reload that Sergeant Thornton knew that he would get only one shot at any force within range.

In fact, when the German attack developed, in keeping with the Britons' greatest fears, it was tank led. Thornton fired his single shot. He had trained well. That single round found its mark, stalling the leading tank and, more important, stalling the entire attack as the German tank convoy backed up like a freeway at rush hour. As a bonus, Thornton's round set off the tank's ammunition supply. The resulting thunder and fireworks convinced everyone, Allied and German alike, that a major battle was under way. On the German side, this meant waiting to gather strength before attempting to retake Pegasus Bridge. On the Allied side, it meant time to land the Pivot force and establish a real defensive barrier. Ambrose writes, "Sgt. Thornton had pulled off the single most important shot of D-Day, because the Germans badly needed that road."[6] How *tiny* the Bearing to the Pivot for the entire Normandy invasion: one small unit, one bridge, one man, one shot.

The Anatomy of the Hammer

For the strategic Hammer, the counterpart to the Bearing of the Pivot is the Hammerhead. On an actual Hammer, the Hammerhead is the weighted business end of the tool, in particular, the striking face of the Hammer. The metaphor of Hammer and Hammerhead is fortunate, because

- The head of a Hammer is the heaviest and strongest part and the part whose shape primarily differentiates the types of Hammers: carpenters', engineers', carpet layers', machinists', nail setters', and so forth.
- The handle supports the Hammerhead while vectoring its force to the crucial location.
- The handle's length provides leverage that multiplies the striking force transmitted by the Hammerhead.

In contrast to the hidden, sometimes furtive Pivot, the Hammer of a company's strategy is usually out there for everyone to see. The Hammer is where the force of a company's strategy is focused, and the Hammerhead is the part of the Hammer that transmits that force.

What I'd like to point out, however, is that often there is only a tiny area in which all that force is concentrated in the Hammerhead—an area analogous to the often tiny Bearing on which the defensive Pivot rests. All the mass, momentum, and force are transmitted through that tiny surface area. A model of the relationship among Hammer, Pivot, Hammerhead, and Bearing is shown in Figure 3-2.

Consider a company whose Hammerhead is its sales force—Tupperware, for example. Right now, Tupperware and its famous parties are succeeding greatly in non-U.S. markets, where Tupperware provides a way for women to enter the workforce that circumvents cultural barriers—they're in the home, yet not entirely in the home. In the United States, Tupperware is struggling with the fact that many women are already in the mainstream workforce, which reduces both the attractiveness of the sales opportunity and the size of the target market. It all comes down to the ability of the Tupperware representative to recruit a hostess, then hold the party participants' interest while she presents a new product. Those few words that propel the twin engines of greed and guilt in the right direction are the Hammerhead for Tupperware's strategy.

Figure 3-2. Strategic anatomy has four key parts.

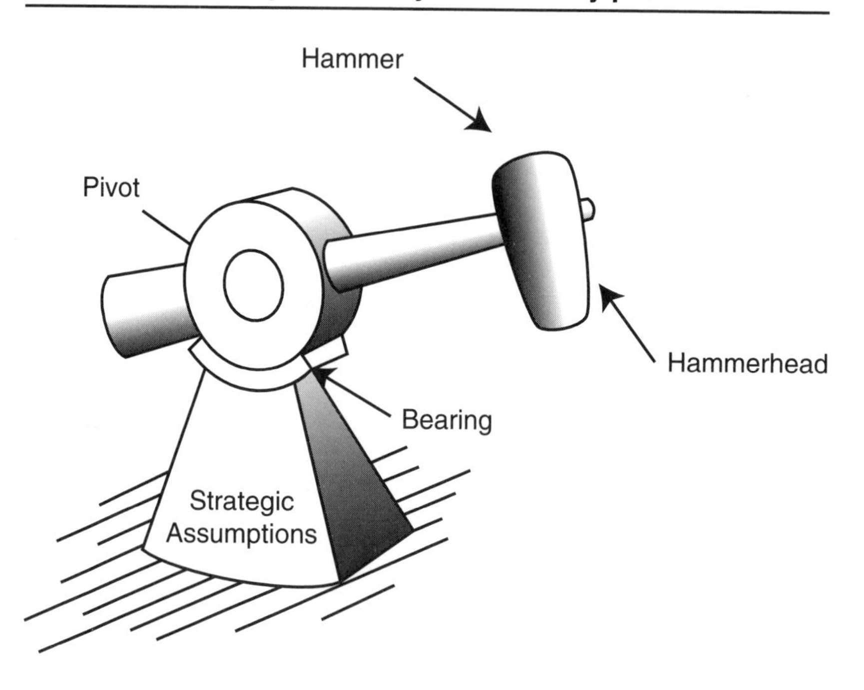

For such sales-oriented strategies, the Hammerhead in the Hammer is pretty easy to recognize. But for other strategies, the Hammerhead is subtler. An example is Kroger, Inc., the huge owner and operator of grocery and general merchandise stores. Recently, Cincinnati-headquartered Kroger embarked on an acquisition adventure, believing that scale economies, especially in purchasing and local market share, are the keys to its survival as smaller grocery chains disappear. Note that Kroger is cash constrained. Its attempts to avoid being taken over in the 1980s drove it deeply into debt. It does, however, operate at a profit today, according to net earnings records from 1994 to 1998.[7] In 1999, it acquired for stock the profitable, growing Fred Meyer chain of food and general merchandise stores in the Northwest. The acquisition will add Fred Meyer's not inconsiderable debt to Kroger's. The financial bet—which I believe should cause some fingernail biting in Cincinnati—is that enough cash can be wrung out of overhead consolidation and purchasing economies to service the debt.

This tells me that the Pivot of Kroger's strategy appears to be the day-to-day ability to operate fairly profitable grocery chains. The Hammer of its strategy appears to be its financial engineering of mergers. And the Hammerhead of the strategy appears to be the ability to sell on Wall Street and Main Street the *promise* of future overhead and purchasing cost reductions to service debt. If we were to dissect a little further, we'd find that crucial to this Hammerhead is a persuasive spreadsheet template or two created by Kroger's finance department and its investment bankers, targeted at the minds of acquirees' shareholders.

Are Kroger's resources concentrated in this very small space? You bet. The investment bankers' fees for the Fred Meyer transaction reached $75 million. Kroger also owed Fred Meyer a fee of $460 million if the negotiations didn't culminate in a deal,[8] and you can bet that 100 percent of top management's attention was on conceiving, negotiating, and consummating the deal—not to mention finding out which top manager was going to have what job when the dust settled.

This example points up two aspects of the anatomy of the Hammerhead:

- It's not always obvious, e.g., where the sales force is blowing its horn.
- Its imprint can be very tiny—easy to miss in your company or your competitor's.

Hammerheads in the Music Industry

A look at the current turmoil in the music industry shows that the anatomy of a Hammer and its Hammerhead can't be taken for granted. The music industry is undergoing a revolution. Record labels are being threatened by the advent of music that can be digitally transmitted and downloaded to personal computers and portable music players. For years, the Hammer of record-label strategy has typically been to field talent scouts to spot acts and artists with potential. Once these are found, the Hammerhead has been the huge dollars spent on publicity, media events, and advertising, all designed to balloon the artist's recognition and create radio station airplay. This creation of demand forces music stores to stock the hot new property, which shoves less well-known artists off store shelves. One result has been the boring lack of variety that cultural critics bewail within music genres. "It's all about scale, baby!"—scale economies in the advertising aimed at the music consumers' minds.

The digital revolution is a direct attack on that Hammerhead. It allows the artists to bypass the record-label publicity machine and get right to the consumers of music. The question is, does anybody care? Established, publicity machine–supported artists are still the draw in most people's minds, and people are still willing to pay for the privilege of hearing them. But they're not willing to pay to listen to anybody else. Current market research suggests that consumers are willing to listen to *emerging* artists only at much lower prices, and that usually means free.

This situation suits most emerging artists just fine, because they know that in their business, fame is the precursor to fortune. But these thousands of unknown and barely known artists need a strategy that neutralizes the Hammerhead of the major record labels—something to make all that money spent on promotion and gaining shelf space irrelevant to a large piece of the music market. The place where they have an advantage is in their local shows. When they are on stage, they have 100 percent market share of the people in that audience. Perhaps an artist can use her presence and name in a local market—say, Laredo, Texas—as a Pivot to launch word-of-mouth campaigns to download her music free (or almost free) from an Internet site. She could allow a fan to e-mail a copy of a track to a friend (a technology that one company, MP3.com, is exploiting now). That could be an interesting way for Sally in Laredo to sidestep the huge Hammer and ubiquitous Hammerheads masterminded in New York, Nashville, and Hollywood.

But the anatomy of a Hammerhead must match the market geography to which it applies. Until a few years ago, the geography of the music business relied on radio play, television advertising and play time, and concert events. The Hammerhead of publicity was designed to maximize a label's presence in those areas—to drive consumers to record stores. But the Internet is changing that geography for the music industry. Theoretically, at least, the Internet circumvents the record label–publicity–music store chain by linking the consumer directly to the artist. (I say "theoretically" because artist quality matters, and because the Internet could easily become clogged with thousands of artists.)

The Internet allows fragmentation of music. It offers a Laredo country music fan access to local artists on a par with access to the big country stars. And the Laredo listener's daughter might have an interest in African music—something never heard on Laredo drive-time radio. Her curiosity can be satisfied by the same number of mouse clicks as required to access the latest crooning from a national country music star. The major record labels' Hammerhead—the heretofore potent cocktail of national publicity, celebrity creation, radio airplay, and hyped concerts—is becoming incongruent with the geography of the market. No wonder there are sweaty palms in their fancy office suites.

The Handle of the Hammer

While our attention is rightly focused on the Hammerhead—the business end of things—we shouldn't overlook the Handle. The Handle is made up of all those resources and functions that work to concentrate force through the Hammerhead.

For example, in the most obvious case, when the Hammerhead is a field sales force, the Handle consists of the functions and resources directly supporting that Hammerhead. In this case, marketing—including sales support, advertising, market research, and pricing—is typically the channel of support for the Hammerhead. Enlisted too, are some not-so-obvious functions such as finance control and research and development (R&D). The former typically supports the credit, creative financing (e.g., leasing, guaranteed buyback) plans, and sales incentive plans the sales force needs to be an effective transmitter of the company's force. The latter stays in touch with customers' demands for new products and features.

When, as in the Kroger example, finance is the Hammer and the Hammerhead is talent to set up deals to meet Wall Street's needs, the Handle is something entirely different. In-house and retained legal counsel are the foremost components of the Handle. Marketing chops back on its work on the latest image-building campaign for its stores. Instead, it gets dragged (kicking and screaming, no doubt) into helping sell the acquisition strategy to shareholders and spin the postacquisition market-share numbers for the benefit of the antitrust watchdogs. Control and purchasing sniff out those often illusory scale savings to be plugged into the investment bankers' pro forma income statements and balance sheets.

Make sense? In reality, this couldn't be more different from the traditional view. In the traditional view, marketing, sales, R&D, and finance are typically equals, at least nominally. It's up to the chief executive to "coordinate" these centers of power and resources to implement the common strategy. There might even be a top management dictate to the effect that, "We are a marketing-driven company." But here we see that once we identify the strategic hypothesis and who or what the Hammer and Hammerhead are going to be, everything else has to subordinate to those. No longer does marketing fashion the strategy and throw it over the wall to R&D and sales to implement.

Yet in another way, as we'll explore in more detail in later chapters, however equal these baronies are *de jure,* most CEOs know *de facto* that one of them is carrying the strategic ball, and, to paraphrase Benjamin Franklin, the other vice presidents had better hang together or they will hang separately.

The Hammerhead and the Wedge

There's a final aspect to the anatomy of the Hammer and Hammerhead that is subtle but important. Whereas the importance of the

Bearing inside the Pivot to the success of a strategy is primarily a matter of the moment—*time*—with regard to the Hammerhead, *location* is as important as time. Let me explain.

Experience suggests that the application of force from the Hammerhead needs to find some success, and that the existence of that success is more important than how great the success is. Part of this is human nature. A strategy belongs to somebody or somebodies; the strategists need to have their vision validated, and as early as possible.

But more important, opportunities for the successful application of force are not arrayed uniformly by nature. Nearly everyone knows that in consumer marketing there are *experimenters* and *early adopters.* The application of force for a new product needs to be targeted at those consumers first, or it will never reach the larger mainstream echelons to follow. No amount of concentrated force is going to get mainstream folks like you and me or laggards (like others we know) to spend a lot of money on the latest and greatest electronic gizmo.

It's an adage in sales that success is as much a matter of numbers and persistence as a good product or a good pitch. You have to knock on a certain number of doors before you get in. But once you're in, you can leverage that success to open other doors, exploiting a referral or a testimonial or simply enjoying the ability to tell new prospects, "I've seen how well this new product has worked for my other customers."

This concept applies to more than just marketing and sales. When a company goes on the acquisition trail, nature seems to spread the opportunities around so that some are more likely to be consummated than others. A failure to complete an initial acquisition appears to dissipate the energy in the Hammerhead and to throw into relief the amounts of time and energy that were wasted in the attempt. A success, however small, seems to renew that energy, as well as to provide the advantage of experience. For a new strategy, the Hammerhead has to *seek out a wedge.* In later chapters, we'll see how this wedge is the cornerstone of making strategic success continuous.

The Hammer and Pivot: Market- or Competition-Centric?

It is a truism that every corporate strategy focuses on both a market and competition. But the reality is that the strategist's attitude—meaning the *company's* attitude—is one of two kinds. It is either *market*-centric or *competition*-centric. In both cases, the concepts of the Hammer and the Pivot illuminate those shadowed but crucial corners we need to inspect closely as we try to develop an open-eyed strategy.

What does it mean to say that a company's thinking is either market-centric or competition-centric? It means that every day the cor pany's chief strategists—everyone who makes decisions about th deployment of resources—wake up in the morning thinking about or of two questions: How can we better find, attract, and serve our cus tomers today? or What are we going to do today to beat our competi tion?

Market-Centrism

Although it is unquestionably myopic to think about serving customers better without thinking about how competitors serve them, and vice versa, it is almost mandatory that one must take priority over the other. This is a natural state of affairs. A company that is based on a new product or a new market or is just plain new *itself* is perforce required to make sure that it has customers and that it is giving them what they want. It's worried about product (or service) definition, appropriate pricing, publicity, and distribution. Usually it's very worried about getting the product out the door.

So you can see it might be apposite to call the market-centric attitude *self*-centric. And although there's something incomplete, there's nothing bad about being market- or self-centric. It's a natural and needed viewpoint for every company in its early stages.

This attitude describes a new Internet-based financial services company I recently worked with. Even though what it sells is as much a commodity as a bushel of corn or a railroad car of coal, senior-, middle-, junior-, and clerk-level attention is all on the product. The reason is that it simply can't take those basic business functions for granted. Every day is an adventure in making sure that the paperwork to get the loans it sells is originated ("sold," to you and me), underwritten, and closed. Ten to 15 percent—tops—of senior management's attention is devoted to the snarls and yelps of the pursuing competition.

When a company is pioneering a new market with a newly invented product or service, attention paid to competition recedes even further. Often these companies feel that they have no direct competition, and perhaps they are right. A medical diagnostic company I worked with enjoyed robust patent protection and lulled itself into believing that its only competition was for the time and attention of the notoriously conservative medical doctors it needed to convince to give its newest test a whirl.

There's another reason why simply being new forces a focus on the internal Hammer and Pivot. It has to do with the pace of change. When a company is new and small, much can go wrong (or right) quickly, with seismic impact. Since the company is small, signing one important new customer might boost sales 15 or 50 percent. That

:an required an overhaul in organization and operations. me company grows to ten times its original size, a new ning calls for a celebration dinner and a change in the ssion spreadsheet, but not a corporate overhaul.

neantime, the likelihood is that changes in the competitive ıre comparatively slower and take longer to have an impact. ı a competitor announces a better product or a lower price, ly have time to react. When I ran my own manufacturing , it was essential for me to react quickly to customer needs, y breakdown, and personnel problems. When I learned that eeled foreign competitor had set up a Midwestern plant with imes the investment and thirty times the capacity of my opera- new that it would take time for Mr. Deutsche Mark to find and t my customers. Nevertheless, with the strategic handwriting on ıll, I put the gears in motion to sell out.

t's an entirely different ball game when the industries involved ıg and mature. Coke spends most of its time thinking about Pepsi, n about BP and Shell, Burger King about McDonald's. New prod- are a self-conscious attempt to grab a fleeting advantage or a des- ate lunge to match the latest product novelties of the competition. ırketing is primarily about the format, timing, and expense of the xt promotion, not about product improvement or new product ex- ıoitation.

It's logical, therefore, that strategists in a new company will take care first to identify their own Hammer and Pivot. They might never, in the course of a year, examine the Hammers and Pivots of the company's competition. Of course, such an approach could easily miss the bull's-eye of strategic perfection, and it might even be a glaring, costly error, as we'll see in a moment. But a new company that is not rolling in cash is inevitably forced to Pivot on functions kept on the brink of starvation and swing the Hammer on others—usually sales or, in technology-based companies, engineering. In these companies, the value of the concept of the Hammer and Pivot is to make crystal clear management's choices about what is to be starved and what isn't. (Extravagantly funded startups and the lucky few that gush cash from day one aren't likely to have to starve any function, of course.)

Using the Hammer and Pivot in Competitive Strategy

Every company that survives grows up. And every company that survives makes the transition from market-centrism to competitor-centrism. It's here that the concept of the Hammer and Pivot really pays off. An effective way to develop a *competitive* strategy—a *competition-centric* strategy—is to identify your competition's Hammer, identify your competition's Pivot, and then identify which of the two you are going to attack, where, and how.

It's unfortunate that business competitive strategy lends itself so easily to the belligerent vocabulary of attack, defense, capture, and surrender. The goal of business activity, after all, is to provide value to human beings. Nevertheless, it's a fact that businesses occupy certain geographical, market, and psychological territories. They take up physical spaces in offices, retail locations, and distribution channels. They occupy prominent or not-so-prominent places in customers' and suppliers' minds, as well as on their computer screens. Progress is made largely through what economist Joseph Schumpeter called the "creative destruction" of these territories and positions.[9] All this is a reminder and apology that there is a reason why the vocabulary of belligerence will always be appropriate for discussions of strategy, especially when the strategic mind-set is competition-centric.

A competition-centric mind-set can generate two kinds of strategic offenses: It can develop hypotheses in which the company's Hammer goes head to head with a competitor's Hammer, or hypotheses in which the company attacks the opposition's Pivot—places where the competition has spread itself thin. More subtly, it can attack the very assumptions in which the opposition has placed its trust—assumptions that neither your company nor your competitors may be fully aware of.

A head-to-head, Hammer-to-Hammer strategy is the most common, most obvious, best understood, and least imaginative of the competitive strategies. For example, if we're Merck and we hear that Lilly is expanding its medical professional sales force by 800 anthropology graduates next year, then we'll increase our sales force, too. After all, we'll have to, won't we? We must match Lilly's face-to-face time with the doctors. Maybe we'll use *drama* majors instead—that will be our competitive advantage.

The net result will be that doctors will see even more of Merck and Lilly, their pharmaceutical market shares will hardly budge in relation to each other; the smaller pharmaceutical houses will get squeezed out of the doctors' appointment books, and a lot of anthropology and drama majors will get jobs. And in two years' time, when the sales vice presidents at Merck and Lilly see that average sales per sales rep have dropped 10 percent, there will be a lot of drama and anthropology majors with pharmaceutical sales experience out on the street.

You would be forgiven if you were to suspect a Hammer-to-Hammer mind-set whenever you see competition based primarily on price. Hammer to Hammer can be a great play, however, if you are in a low-cost position: When your Pivot is an unassailable low-cost manufacturing position, your Hammer relies on aggressive pricing.

But that's only the most obvious case of Hammer to Hammer. In such industries as chemicals and oil—and increasingly the auto industry—where products are commodities and pricing is worldwide, Hammer-to-Hammer competition takes the form of investing aggressively to achieve low-cost production. Since this usually means seeking economies of scale, new investment classically leads these industries to overcapacity, followed by price-cutting to fill up that capacity and a shutdown of less-efficient plants or players with weak balance sheets and nauseous stomachs.

These are fairly common, classic examples of Hammer-to-Hammer competition. But Hammer to Hammer also lurks where you might not expect it. It springs up, smiling its tantalizing, wicked smile, when Barnes and Noble puts up a store just down the street from Borders Books; when Democratic and Republican Party strategists compare the minutes of commercial TV airtime their rubber-chicken fund-raisers can support; and when Ford, GM, and Chrysler announce 1.9 percent car financing within days of one another.

If you were to hold up the looking glass to your competitors, would you recognize their strategy as your own? If so, it's likely that you've got a Hammer-to-Hammer strategic mind-set.

Attacking Assumptions: Hammer-to-Pivot Strategies

Now let's explore the other competition-centric strategy, called "attack the Pivot." Once we've identified the competitor's Hammer and Pivot and we've shuddered at the costs of going Hammer to Hammer, the next place to look is at the competitor's Pivot. (For simplicity, we'll pretend we have only one major competitor; the ideas can be extended when there are multiple competitors.)

Recall that the competitor's Pivot is where he is playing defense, where he's stretched thin, where he's relying on the robustness of his current position to send resources elsewhere. If it's successful, a competition-centric strategy focused on your competition's Pivot pulls the props out from under his strategy.

A wonderful example is unfolding now in the video rental business. Blockbuster Video has signed deals with the major motion picture studios to obtain exclusivity on movie releases at low or zero cost. Instead, the studios will share in a percentage of movie rental receipts with Blockbuster.[10] Since major motion pictures—the season's hits—are vital revenue sources for all video stores, independent stores are screaming to the Justice Department. Blockbuster is attacking their *air supply:* the ability to get products in their stores.

The ability to obtain movies from studios is what the independents relied on—it was a fundamental part of their Pivot. Their Hammer was location—a convenient storefront and service. A Hammer-to-Hammer strategy by Blockbuster would be to outlocate the independents by opening more and more stores. And no doubt this is part of what it's doing. Nevertheless, those pesky mom-and-pop independents just spring up anywhere they can rent a few thousand square feet and nibble into Blockbuster's high-cost, high-selection offering.

Another example of an attack on a competitor's Pivot comes from the world of the might-have-beens. Toyota's export drive in the United States began in the 1970s, with its first success coming in the wake of the oil embargo in 1973–74. Toyota's Pivot was its extraordinarily strong position in the Japanese domestic market. That market provided the supplier network, manufacturing scale, and cash to fund the export drive. A Hammer-to-Hammer response by the U.S. manufacturers would have been the development of their own import-beating compact cars. And that's exactly the road they chose. You may remember the Ford Maverick, the Chevy Monza, and the AMC Gremlin. But you will be forgiven if you don't: These were not among Detroit's most memorable nameplates.

A Hammer-to-Pivot strategy would have been quite different. It would have focused on limiting Toyota's sales in the Japanese domestic market. Perhaps a concerted export drive, combined with lobbying to loosen Japanese import tariffs and nontariff import barriers, would have given Toyota something to worry about besides what Californians thought was cool. (See Figure 3-3 for a schematic conception of such unconventional competitive thrusts.) Or—perhaps more pragmatically—aggressive building of ownership stakes in Toyota's domestic competitors, combined with funding of their sales campaigns in Japan, might have redirected Toyota's attention to the home front. That such major stakeholding was feasible is proved by GM's stake in Suzuki, Chrysler's stake in Mitsubishi, and Ford's stake in Mazda. It appears that these stakes never translated into a competitive strategy as such, however. They led only to alliances for building smaller cars, sourcing components, and establishing raw financial investments—or, in the case of the Mazda stake, a turnaround headache for Ford.

A third competition-centric strategy is predicated on attacking your competitor's *center of gravity*. It deserves mention here because it fits nicely into the logic of strategies based on seeking out competitors' Pivots (see Figure 3-4). To understand this, we have to understand the concept of a center of gravity. It's borrowed from Clausewitz, the military theoretician. Its evolution is a remarkably concise statement of the

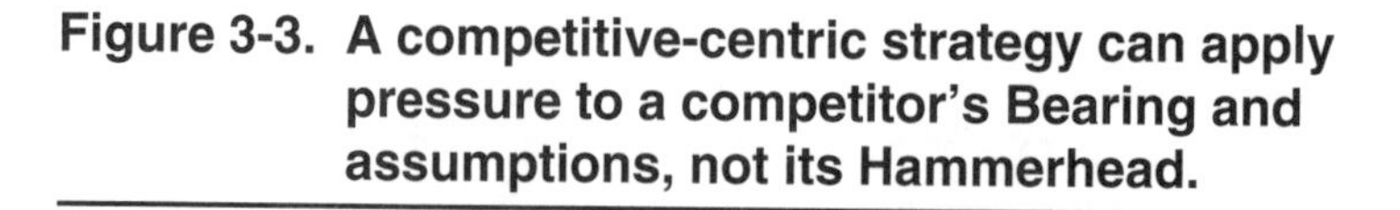

Figure 3-3. A competitive-centric strategy can apply pressure to a competitor's Bearing and assumptions, not its Hammerhead.

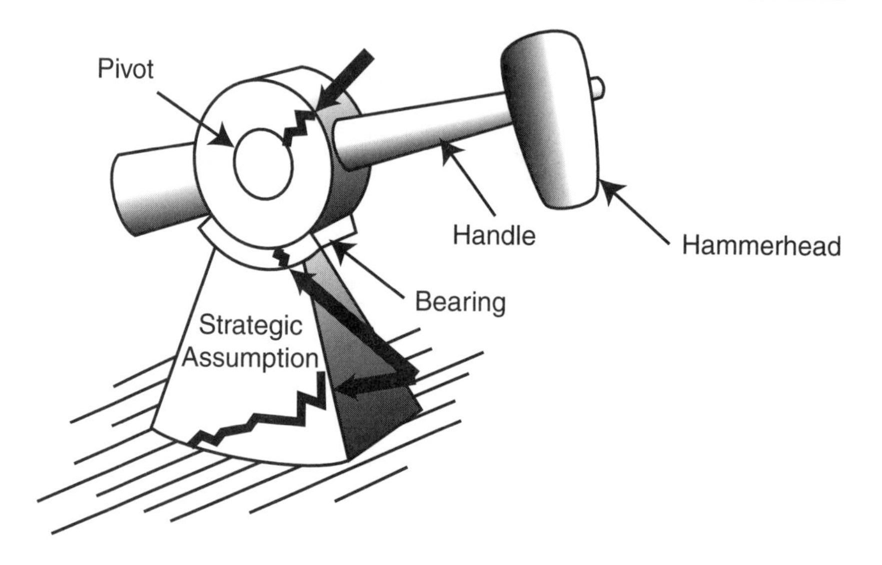

Figure 3-4. Competition-centric strategies offer thought-provoking choices.

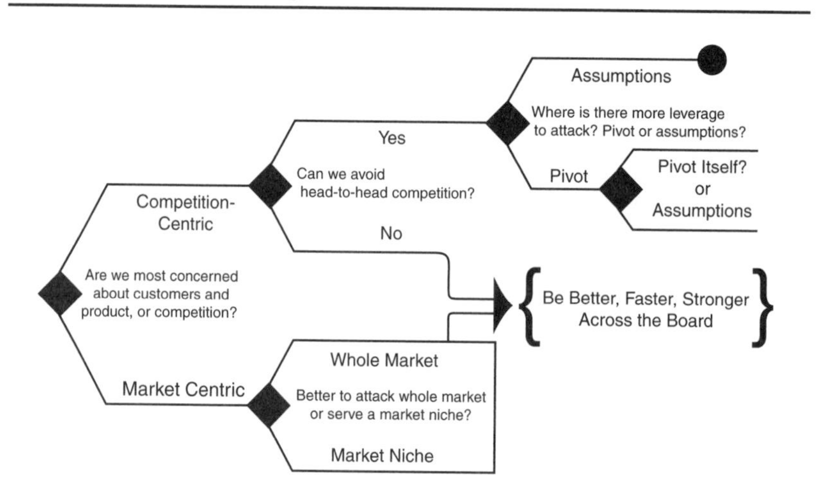

difference between a Hammer-to-Hammer strategy and a Hammer-to-Pivot strategy. The official U.S. Marine Corps warfare manual states:

> There is a danger in using his term [center of gravity]. Introducing the term into the theory of war, Clausewitz wrote (p. 485 of *On War*): "A center of gravity is always found where the mass [of the enemy] is concentrated most densely. It presents the most effective target for a blow; furthermore, the heaviest blow is that struck by the center of gravity." Clearly Clausewitz was advocating a climactic test of strength against strength. . . . But we have since come to prefer pitting strength against weakness. Applying the term to modern warfare, we must make it clear that by the *enemy's center of gravity we do not mean a source of strength,* but rather a critical vulnerability.[11]

Clearly, what Blockbuster has done is target a critical vulnerability in the independent's Pivot, and it's something that an individual store can do very little about. This suggests that the appropriate response for the independents is to band together in a buying-power cooperative.

In the 1970s and 1980s, Pepsi found a critical vulnerability in Coke's broad demographic appeal. It used advertising to portray itself as the soft drink for the *young*. By extension, of course, Coke was seen as the drink of the unappealing old.

Microsoft's source of strength is its twin dominance in office software and in operating systems for personal computers. Certainly as a competing software company you could go Hammer to Hammer with Microsoft, offering a better word-processing program (many argue that WordPerfect *is* one) or a better Internet browser (certainly Netscape's browser is the equal of Microsoft's if not superior). Or you could go after Microsoft's critical point of vulnerability—the quality of its products. It would be interesting to see how a product and a campaign based on bulletproof reliability would fare. And there's that other critical point of vulnerability: the very market share that makes life such a challenge for its competitors—the antitrust laws.

An attack on a Pivot that is a vulnerable center of gravity may be the wisest of the strategic alternatives. And it may be less dangerous than an attack on the most blatant aspect of a competitor's Pivot. First, a Pivot is in defensive mode, which is a stronger position than offense. Second, an attack on the Pivot is sure to be noticed and draw resources from your opponent's other projects. For example, in the 1990s, low-priced upstart airlines such as Western Pacific, Sun Jet International, and Vanguard Airlines began offering discount fares from Dallas–Fort

Worth (DFW) airport to Wichita, Kansas; Kansas City, Missouri; and Colorado Springs, Colorado. They were, of course, attacking the second largest airline in the United States—American Airlines. And they challenged its dominance at its home base and main terminal at DFW. Internal American Airlines documents show that it wasted no time in slashing fares and adding flights to protect its core operation, its Pivot. Today, two of the three upstarts are gone from the market, American fares that had dropped 20 to 40 percent are back to precompetition levels, and passenger count has dropped. According to an antitrust lawsuit filed by the U.S. Justice Department, American Airlines knew that it would lose money by cutting fares and adding capacity: "The short-term cost, or impact on revenue can be viewed as the investment necessary to achieve the desired effect on market share."[12]

These upstart airlines attacked American's Pivot but overlooked three factors. First, since DFW was American's primary hub and its headquarters, any attack would be sure to gain its immediate attention. Second, this was a battle that American Airlines could not afford to lose: DFW was its flagship airport. Third, American had the bulk of its assets and resources at DFW. Whereas Clausewitz would recommend a climactic battle over DFW, the marines' interpretation would be that the upstarts were attacking a concentration of strength—something to be avoided, if possible. What they were not doing was attacking a *crucial vulnerability.*

Of the three upstart airlines, Western Pacific was the first to go bankrupt; the other two are still operating in other markets. Western Pacific had the temerity to take on not only American at its hub but also industry giant United Airlines at one of its main hubs, Denver.

Spying out your competitor's center of gravity and key vulnerability is a creative art. It demands an understanding of your competitor's *big picture,* as well as the ability to isolate any instance of a Bearing inside your competitor's Pivot. An example of big-picture thinking is Microsoft's battle with Netscape in the 1990s over the Internet browser market. At the time of Microsoft's market entry, Netscape had over 80 percent market share. Microsoft, waking to the fact that the new marketplace was the Internet and that the browser was the gateway to that market, realized that Netscape's crucial vulnerability was not its product (which was clearly superior to Microsoft's first browser efforts), not its programming (ditto), and not its brand name (which was widely recognized) or its finances (Netscape had plenty of cash), but the very fact that Netscape had gone public at a high stock market valuation.

To keep that valuation up—and thus sustain the implicit promise of spectacular future success that a high market valuation carries with it—Netscape needed to show the capital markets success every quarter, either in revenue or in market share, and preferably in both. Microsoft, contrariwise, made a habit of downplaying its quarterly profit forecasts to stock market analysts. It deftly played the game of reducing Wall Street's expectations by highlighting the competition's successes and stashing away earnings in accounting reserves. That gave it breathing room to take an earnings hit if it needed to. When Microsoft launched its browser for the low price of $0, it wasn't just a price war. To satisfy the Netscape investors' cravings for the prospect of share and earnings, Netscape had to respond to Microsoft's move by turning away from the consumer browser market and toward the corporate Internet business. By 1998, Netscape was forced to match Microsoft's browser price, and in 1999, the two have rough parity in browser market share. Who would have thought that going public at high valuation, and raising lots of cash in the process, could be a source of critical vulnerability rather than strength?

Perhaps yours isn't a giant company like Microsoft, or even an aspiring giant like Netscape. Searching for the point of critical vulnerability in a company's Pivot can be rewarding in the smaller leagues, too. A professional services company I know wound up in a dispute with a customer over work it said it had performed admirably and the customer claimed had hardly been started. The services company automatically set in motion the conventional series of strident demands and threats to send the overdue bill to collection.

But a chat with a savvy outsider convinced the company of the superfluity of this approach. The customers' area of critical vulnerability, he saw, was its reputation. The customer had received press for a new technology all out of proportion to either its size or its success. It hoped to go public in a few months. Its venture capital backers were nervous about the true prospects of the company's success and wanted to get the initial public offering out soon.

Creative action. A carefully worded call to the customer's public relations counsel hinted that the customer's already fragile reputation might be damaged if it became known that it didn't pay its bills.

Follow-up action. A similar call was made to the lead venture capitalist's partner in charge of the customer.

Result: payment in full.

Strategy as Hypothesis Revisited

In chapter 2, I suggested that a strategy is a hypothesis consisting of an *if-then* statement. The *if* clause sets out the conditions under which you expect certain (favorable) outcomes. So far so good. Our examples—Blockbuster versus independents, American Airlines versus upstarts, Microsoft versus Netscape—suggest that the search for a competitor's vulnerable point, the center of gravity for the attack, should not overlook the *unstated assumptions* of the opponent's strategy. This leads us to a more complete model of a strategic hypothesis:

1. **Given certain key assumptions (about consumer behavior or competitors or macroeconomic conditions or the regulatory environment) . . .**
2. **If we create explicit conditions a, b, c . . .**
3. **Then we can expect results x, y, and z.**

It is the creative search for these assumptions—so obvious as to be unconscious—that gives rise to the most compelling competition-centric strategic hypotheses.

In physics, the stated and unstated assumptions include properties of mathematics, underlying laws of nature, or the theoretical framework—Newtonian, relativistic, or quantum-mechanical. In business, the unstated assumptions most frequently deal with the capital market or the financial, legal-regulatory, and public-opinion environments you and your competitors are operating in. Assumptions that go under the rubric of *industry tradition* are sometimes stated and sometimes unstated. As we saw with Blockbuster, an assumption the independent video dealers held about the business was that they could always bid for movies against any competitor. Probably no one thought about a different way to source movies until Blockbuster hatched the exclusivity and revenue-sharing scheme with the major studios. That took creativity and insight. And there was no royal road to it.

Searching out business assumptions and using money, technology, or sheer brains to invalidate them goes by the familiar catchphrases *changing the game* and *judo strategy*. No one can give you a road map to them. But there is a fact-based process you can go through to help turn the lights on these hidden assumptions. We'll get into that in the next section.

For Tomorrow Morning: Identifying Your Hammer, Pivot, and Bearing, and Choosing a Market- or Competition-Centric Strategy

By now, you probably have a good idea where your company's current Pivot and Hammer reside. Perhaps you know, too, what the crucial Bearings and Hammerheads are. Perhaps you even see that your competitive strategy—at least toward your company's main rivals—is regretfully Hammer to Hammer. But how would you lead a meeting tomorrow to gain the agreement of your colleagues on these critical identities? How do you bring to their awareness the importance of strategic choices already made or in the making?

Here's one way.

Suppose your strategic thinking is in market-centric mode. Start by making a simple chart of your company's business system. This would typically include all the major functions needed to create a product or service, deliver it to a customer, and get paid for it: R&D and product development; engineering, manufacturing or operations, marketing, sales, distribution, service, and finance and control; plus crucial overhead functions such as human resources and legal and regulatory functions.

This is the familiar *business system* of consulting parlance. However, it is best to err on the side of inclusiveness when you are seeking Pivots, Bearings, and Hammers. Often, vulnerable Bearings—yours or your competitors'—are hidden in easily dismissed "merely overhead" functions such as human resources and customer credit.

Now, alone or in a group meeting, decide in which of these areas you are playing offense or defense. Within the marketing function, for example, you may be playing defense with some products or services and offense with others. In the department store example at the beginning of this chapter, we saw that senior management assigned the flagship retail business a defensive role while it went after new catalog and overseas business.

Suppose you are general manager of RadGen, once a technology and market-share leader. But now, competition has caught up with you in technology, and it's gaining on you in market share. You've decided that without a new generation of products a quantum jump ahead of the competition, your future is far from rosy. At the same time, you have a loyal customer base generating ample, steady service revenues. Your chart might look something like Figure 3-5.

The next step is to identify the Bearings and the Hammerheads within each of RadGen's key areas designated for offense and defense. Let's truncate the chart to just the key Pivot and Hammer functions, so it doesn't get unwieldy. The key here is to ask where, who, and what the details are about those Bearings and Hammerheads—to drill down vigorously, as we saw in our examples. Try creating a chart like Figure 3-6.

The reward for drilling down so far is the crystal clarity you get about where to expend resources you've been hoarding. Look at Figure 3-7.

Figure 3-5. RadGen chooses to play offense in R&D, engineering, and marketing and defense everywhere else.

R&D/product development	*Engineering*	*Manufacturing/operations, including purchasing and sourcing*	*Marketing, including pricing, public relations, market research*	*Sales, including channel relationships*	*Service, including customer service, field service, customer care*	*Finance and control, including relations with capital markets, investors, and banks*	*Human resources, including recruitment, training, managing "alumni"*	*Legal and regulatory, including lobbying federal, state, international, and standards organizations*
Offense: Focus on the "Apollo" project Defense on other projects	Offense: Liaison with R&D Defense on remainder	Defense: Focus on cost reduction awaiting new products	Offense: Team assigned to R&D project Majority, defense: Focus on current major account retention	Defense: Focus on current customers; switch temporarily to promoting service contracts	Defense: Focus on very good service to existing customer base	Defense, except team members assigned to R&D project to fast-track new product cost of goods Offense with regard to hyping "Apollo" on the street after 5/17	Defense, including allocating human resources to R&D project	Defense, except offense on lobbying efforts needed to obtain federal approval of new product and favorable industry standards

Figure 3-6. A company seeks out its Hammers, Pivots, and Bearings.

R&D	*Sales*	*Service*
The Hammer is: Apollo project team, Dr. Lee Wing, project manager	*The Pivot in Sales:* Our sales force, Elsinore Singer, vice president	*The Pivot in Service:* Our field service team, Cory Norhoff, director.
The Hammerhead is: Dr. Wing's experience in developing light-sensitive organic crystals. Most important are three crucial experiments he and his recently recruited coresearcher, Dr. Candace Christian, are running in Culver City, California, and Lucerne, Switzerland. Wing's demands that Christian fly every week to Lucerne by way of Austin, Texas, may be endangering commercialization of experiment results.	*The Sales Bearing:* James Richardson, branch manager, Lexington, Massachusetts. Has relationship with Syntactic Displays, a key account, says key to keeping the account is "Red Sox baseball tickets." Second most important account: Sung Moon Video Division, Rhee Kim, general manager. *Stability of account unknown: Local agent in Seoul not responsive—need to find out.*	*The Service Bearing:* Senior, long-term people serving largest, most important accounts (Syntactic and Sung Moon). Other key accounts important to Bearing dependent on recent college trainees led by Jean Storch. Storch is bright and high energy, so Bearing here is Storch's *leadership and confidence skills.* Norhoff says he is spread too thin to see Storch more than once a month and improve Storch's coaching styles.

Figure 3-7. Steps for allocating strategic resources.

R&D	*Sales*	*Service*
Identify understudy or European-based research team cocaptain to ride herd on Lucerne experiments and reduce Christian's travel. Find way to document what Wing knows. Find way to reward Wing and Christian for working together and completing experiment successfully, while being careful of researcher egos.	Define clearly for Singer and Richardson what constitutes success with respect to Syntactic and Sung Moon during the Pivot period. Hold workshop with Singer and Richardson to reallocate some sales force resources to better cover Syntactic and Moon. Get Singer to Korea fast. Find new agent or recruit local account rep. Meet with Kim and check Moon's financial condition and customer relationships.	Review (outdated?) service standards for key accounts with Norhoff. Revise so they apply to all key accounts. Perform ride-along with Storch and several new reps. Develop coaching plan with Norhoff for Storch. Possibly promote one of Norhoff's better field service engineers to be a second in command for Norhoff or get Norhoff some administrative help to free up more road coaching time.

Developing a Competition-Centric Strategy

Are you ready to take the leap to a competition-centric strategy? Again, there's no royal road to a brilliant competitive strategy. But a look at your competitor's business system, its key assumptions, and where its Pivots and Hammers are positioned can put you in the driver's seat, instead of reacting to your competitor's moves.

Here's how you, as founder of Silver Star Industries, a market-share follower but recently a technology peer of RadGen, might start the pot boiling for competitive thinking. Let's assume that Silver Star knows nothing about RadGen's decision to funnel resources into Wing's research while going on the defensive elsewhere and relying on a key account sales and service Pivot. You think that your technology is equivalent to RadGen's, and you know that you've gained share over the last year. You may be right, you may be wrong, or you just may not know what you *need* to know about your main competition. But a sharp look at RadGen's business system in Figure 3-8, pages 80–81, might find a tiny crack in its dam of seeming invincibility.

There's a tremendous payoff for Silver Star from all this drilling into RadGen's vulnerabilities. We can create at least two potentially testable, falsifiable strategic hypotheses for Silver Star—one focused on major accounts, and the other on secondary accounts.

Strategy A

Given RadGen's apparent vulnerability with the etching tool vendor and the value-added seller, *if* Silver Star creates a closer partnership with either or both, while revamping its sales message and forestalling any new, unfavorable standards, *then* it can crack open at least one major account and open the doors to other major and secondary accounts with superior delivery and possibly with better pricing.

Once we do our homework to define the potential deal outlines and (therefore) the costs of these partnerships, we can test the waters with the potential accounts. We can survey them, using an independent firm to see whether improved delivery (made possible by our relationship with the etching tool vendor) and a switch in value-added dealer allegiance, plus our improved technology, will be enough to sway their purchase decisions to our side of the ledger. Silver Star might well be able to perform this test *without* actually entering into those agreements with the etching tool vendor and the value-added dealer.

Strategy B

The other high-potential strategy based on this analysis is something like this:

Given Silver Star's apparent focus on large accounts, *if* we focus our sales, new recruiting, and service resources on secondary accounts while delivering a carefully crafted message to Wall Street, *then* we will generate enough increased sales of our own to warrant the effort and to put a lid on RadGen's market share, thus choking its capital supply.

This, too, leads directly to a testable hypothesis. An outside firm can find out how many secondary customers we're likely to convert to our product. A little math will show whether that would indeed put a limit on RadGen's market share and earnings and significantly affect its stock price.

How beautiful to see the Pivots and Bearings and Hammers and Hammerheads pop out in strategies like these!

Strategy A

Hammer: New relationships with tool vendor and value-added dealer, leading to industry-leading service
Hammerhead: Major account sales
Pivot: Our R&D, engineering, and other marketing and sales efforts
Bearing: Our ability to maintain technological parity during the attack

Strategy B

Hammer: New secondary-account sales and service
Hammerhead: Veteran salespeople switched to secondary accounts, leading specially picked new sales recruits
Pivot: Our R&D, engineering, and other marketing and sales efforts
Bearing: Our ability to maintain technological parity during the attack, and especially our ability to influence the technology standards process

---

You can use the template in Figure 3-9, similar to Figure 3-8, to help find the main vulnerabilities in your leading competitors' business systems. You'll need one for each competitor. And then you'll have to develop strategies like Strategies A and B to get you started on creating a competitive strategic hypothesis. Take a crack at filling out Figure 3-9 (you'll have to fill in the actual names of your competitor's business system functions in the left-hand column—I've given you generic names as thought-starters).

Figure 3-8. Developing a competition-centric strategy against RadGen.

RadGen's Business System Function	*Where Might RadGen Be Vulnerable?*	*Best Place to Attack*	*Overall Potential as a Competitive Strategy*	*Employing What Competitive Means*
R&D/product development	Current patents Hires lower-level researchers	Legal challenge to patents Faster environment for researchers	Patent challenge: medium—raises industry doubts on RadGen technology Researcher hiring: high value, long term	Tout better working environment, more challenge, same money to researchers
Engineering	Hardly at all; industry's best engineering team	I give up	Poor; we must maintain parity but not go head to head	Keep parity by cherry-picking a few very good engineers, good environment
Manufacturing/ operations	Industry dependent on a single etching tool supplier; RadGen has beat it up on price	Etching supplier must make large investment in new tooling, leaving it even more vulnerable to RadGen	Very high; potentially coinvest with etching tool supplier in return for faster delivery at lower price	Open negotiations now!
Marketing	RadGen can no longer tout industry technology leadership	Medium and large customers afraid of being dependent on RadGen	Low to medium, but with low cost and long-term high payoff	Revamp sales literature to show technology parity; change sales force pitch

Figure 3-8, cont'd.

Sales, distribution	Best front-line sales force, but RadGen has beat up third-party value-added OEM suppliers	Major RadGen customer, Sung Moon, dependent on key RadGen value-added supplier who is tired of RadGen's business practices	Very high	Capturing allegiance of Sung Moon's value-added supplier and others could crack larger accounts; open discussions now!
Service, field service	Service to major customers is great Service to second customers: not so great	We could overserve customers who are second-tier to RadGen, making them defect	Moderate to high, if we focus our best salespeople on second-tier accounts instead of trying to break large accounts	We need to pencil the economics of this out Main problem: large account mind-set of our best salespeople
Finance and control	Lowest cost producer Stock price is at 30-week low	Picking off key second-tier but highly visible customers will keep RadGen's stock price down, encourage defections	Low to moderate	Beef up our contacts with Wall Street analysts, sowing seeds of doubt about RadGen
Human resources	Need constant supply of fresh graduates; high salary, but known as a "clone dome"	Recruit up-and-coming RadGen researchers, salespeople; create disruption	Moderate—could be crucial to a second-tier strategy	Need more explicit "competitive recruitment plan" Our current human resources vice president, from RadGen, needs reeducation
Legal and regulatory	RadGen suddenly pushing for closure on new industry standards Why?	Standards association meetings; need to use our sales and R&D people to explain RadGen behavior	Very high at least in short term; a regulatory or standards barrier could buy us 2–3 years' time to find out what RadGen is up to	Need a combination R&D, regulatory, and lobbyist task force to figure this one out

Figure 3-9. Template for developing competition-centric strategies.

Name of Competitor: ____________________

Leading Competitor's Business System Function	*Where Might Competitor Be Vulnerable?*	*Best Place to Attack*	*Overall Potential as a Competitive Strategy*	*Employing What competitive means*
R&D/product or service development; intellectual property rights				
Engineering/product or service development, including concepts in process				
Manufacturing/operations, including proprietary skills, processes, and process economies; operational mind-sets and traditions; includes key supplier relationships				
Marketing, including "hard" marketing assets, reputation; relationships				
Sales, distribution; channel needs and expectations				
Service, field service; including relations with third-party sales and field service organizations, including dealers				
Finance and control/capital structure and source of funds; investor relationships and expectations				
Human resources: key people, including leadership; compensation, training and promotion policies; working environment				
Legal and regulatory; political/lobbying situation				

Notes

1. For an entertaining discussion of the dangers and evils of brand extension, see Al Ries and Jack Trout, *Marketing Warfare* (New York: New American Library, 1986), 117–38.
2. Of course, large companies usually have multiple business units with correspondingly multiple strategies. The above observation applies to strategies for a single business unit.
3. David S. Landis, *The Poverty of Nations: Why Some Are So Rich and Some So Poor* (New York: W. W. Norton, 1998).
4. Dava Sobel, *Longitude: The True Story of a Lone Genius Who Solved the Greatest Scientific Problem of His Time* (New York: Walker and Company, 1995).
5. Stephen Ambrose, *Pegasus Bridge* (New York: Simon and Schuster, 1985).
6. Ibid., 119.
7. "Form 10-K: The Kroger Co.," Securities and Exchange Commission, 2 January 1999.
8. "Form 424B3: Joint Proxy Statement Prospectus (The Kroger Co./ Fred Meyer, Inc.)," Securities and Exchange Commission, 9 March 1999, 10, 20.
9. Joseph A. Schumpeter, *Capitalism, Socialism and Democracy* (New York: Harper and Row, 1950), 81.
10. David Segal, "Fast-Forward Deals: Blockbuster Thrives on Arrangements with Studios," *Washington Post*, 15 September 1998, C1.
11. Quoted in Robert Leonhard, *The Art of Maneuver* (Novato, CA: Presidio Press, 1991), 21.
12. *New York Times*, 16 May 1999, A1.

Chapter Four

The End of an Era: The Twilight of Sustainable Competitive Advantage

> We have good reason to believe . . . that changes in the condition of life give a tendency to increased variability.
>
> —CHARLES DARWIN

In the last chapter, we saw how Silver Star Industries, RadGen's aggressive, smaller competitor, might have created a competition-centric strategy to put pressure on its larger, established rival. Our unexpected conclusion was that perhaps Silver Star shouldn't be expending energy on building a better mousetrap—a better product—at all. Rather, perhaps it should heat up the battle with RadGen on points that might *never* come up in a sales brochure or an annual report, or even in an economic value-added analysis from a major management consulting firm or one of Wall Street's finest investment banks. Perhaps it should forge an alliance with a crucial tool vendor, spook the industry's technical standards committee, or use a backdoor approach to one of RadGen's biggest customers. Perhaps it should do all or any of these rather than trusting to a day-to-day, salesperson-to-salesperson slugging match on product features, performance, and after-sales service.

What Silver Star is seeking, of course, is an *advantage* over RadGen. This chapter dissects the idea of advantage—what kinds there are, and where they are to be found. Importantly, we will skewer one of the mainstays of strategic thinking for the last generation—the idea that strategic position consists in finding a sustainable competitive advantage (SCA). SCA, we'll discover, is so wounded that it lives only in remote locations in the modern industrial jungle. Fortunately, SCA is survived by a new, fast-growing offspring, opportunity creation and exploitation (OCE), which is described in the next chapter. If your company's senior managers believe that your success depends on finding and keeping an SCA, you will find this chapter and the next one particularly provocative.

Baseball Has Been Very, Very Good to Me—But Not to Strategy

The best way to approach our quarry, the concept of competitive advantage, is to sneak up on it almost as a physicist would: by finding a place where it can be isolated and, for all its ambiguous squirming, pinned down. One place where we can pin down the competitive advantage concept is in a study of its relationship to strategy in American baseball.

A favorite analogy among business commentators is that business is like baseball. When it comes to teamwork, skills, and communication, I'll admit that there's a fair amount of truth to the analogy. But nothing could be further from the truth when it comes to using the game to model competitive strategy. In fact, using this analogy can be downright detrimental to energetic strategic thinking.

First, if it makes sense to talk about the *design* of baseball, we might say that it is designed as a team sport that paradoxically does everything it can to highlight the talent of individuals. It is primarily a private duel between batter and pitcher. When a hit is struck, it is usually a contest between the runner and one or two basemen. Add to this the fact that the players' positions are almost entirely fixed. There's no opportunity to concentrate resources, outguess an opponent, or take a calculated risk at an inning's start—all hallmarks of strategic thinking for companies. The principal opportunity for any maneuver comes in the pitcher's choice of what kind of pitch to throw and the fielder's choice about whether to go for one out or a double or triple play.

This purity is what makes baseball fascinating. Like a beautiful physics experiment, it narrows the variables down to a few human actions that, if done slightly differently, can spell the difference between victory and defeat. It highlights the sharply defined features of singular, idiosyncratic athletic prowess. And it explains the fans' fascination with the players' personae and personalities.

Second, teams compete on the same playing field—not just different parts of the same field as in football, soccer, or tennis, but the very same field. Companies, however, all have unique histories, and those histories ensure that they occupy different spots in the marketplace in terms of real estate or customer relations or distribution channels. Despite the best efforts of lawyers and antimonopoly officials, the corporate playing field is not only never level, it isn't even the same size. But it is just these differences that make all the difference in what strategies are feasible for a company and which ones are not.

Third, baseball teams are blessed with practically perfect intelligence—not the IQ kind, but knowledge of the other team's dispositions and behavior. *Dispositions* are known, because players' posts are pretty much set by custom, rules, and the necessity to keep the entire field covered. *Behavior* is known, because rules tightly regulate what's allowed and what's not; what cannot be regulated—sheer talent—can be inferred by watching future opponents in other games. By contrast, in the last chapter, Silver Star could only infer RadGen's intentions to launch a new generation of products by its peculiar behavior at industry meetings on standards.

Fourth, each baseball team's total resources in any given game are exactly the same—the total number of players (leaving aside injuries and other unexpected circumstances). Part of the beauty of the game is that it highlights the singular differences in individual human capability. Two companies almost never have strictly comparable resources, either in kind or in quantity. RadGen is larger, better capitalized, and has more salespeople than Silver Star.

Such strategy as exists in baseball consists in deciding which batters to play and which to bench. In professional baseball, it's strategic to decide preseason which ones to keep and which to trade, but that's as far as it goes. Next time you're tempted to compare your company to your favorite baseball team, stick to comments about individual performance and tactical teamwork.

The Difference that Difference Makes

My purpose in laying out what is perfectly plain is not merely to warn you about taking sports metaphors too closely to heart in business. Rather, it's to illustrate that the fundamental raw material for corporate strategy is *difference*. Baseball in particular—but other sports, too—tries so hard to eliminate difference that it eliminates strategy.

The real business lesson I've gathered from baseball is that when you eliminate difference you shrink the scope for strategy. Let's look at an interesting example from the food industry to see what a difference *difference* makes.

In March 1997, McDonald's was facing declining same-store sales and stiff competition from other fast-food chains. It announced plans to cut prices radically on its flagship Big Mac and Egg McMuffin products to fifty-five cents. The apparent reason: McDonald's could think of no better way to be different from Burger King, Wendy's, and Burgerville—and even from Taco Bell and Arby's—than *price*. One industry analyst said that McDonald's had "transformed one of the great brands in American business into a commodity."[1]

Perhaps McDonald's was betting that Burger King and Wendy's would leave the business when they realized that they couldn't compete against McDonald's lower prices, pure and simple. But if this was the case, McDonald's soon realized that the bet was lost. Consumers saw that there wasn't any real saving involved, because the new low prices were available only when the flagship products were purchased as part of bundled meals. And the other fast-food chains quickly retaliated with their own "value meals."

Sensing that it still makes sense to be different, McDonald's has tried several paths. In 1996, its Arch Deluxe product introduction failed to lure adults in adequate numbers, so in 1997, McDonald's moved back to its tried-and-true strategy of focusing on kids, and it has beefed up its relationships with Disney and other movie studios for promotion tie-ins. By 1999, McDonald's was moving back up in market share, but the fifty-five-cent strategy had nothing to do with it. The difference is no longer in the food or the pricing but in the tie-ins with the entertainment industry—food redefined as a form of entertainment rather than nourishment.

Compare this story with another from the food industry, that of Snapple, which makes bottled fruit juice–based drinks with offbeat names like Whippersnapple and Frapple. Snapple was founded in 1972 by independent entrepreneurs Arnold Greenberg, Leonard Marsh, and Hyman Golden, not as the product line of a larger food processor. At first, Snapple drinks were distributed in guerrilla fashion to small outlets such as gas stations and convenience stores. It had quirky advertising and a unique taste designed to appeal to a limited, albeit loyal, flavored-tea or juice-cocktail market segment. These characteristics help explain why Snapple was highly successful as an independent brand but spectacular and costly failure as part of Quaker Oats.

Quaker paid a pretty price for Snapple ($1.7 billion) in 1994.[2] Obviously, Quaker bought the brand with the idea that its big-bucks marketing budget and access to shelf space in America's grocery stores could multiply sales. Quaker set out to make Snapple appealing to mainstream America by broadening its taste appeal and mellowing its image. The flavors lost their distinctive edge, and the ads lost their unique flavor, too. Quaker tried to use Snapple as a way to take on Coke and Pepsi on their home turf.

The result was disaster. Although Snapple had made money for years (flavored water has a low cost of goods) as an independent company, Quaker lost $6,600 an hour on Snapple.[3] It wound up auctioning off the brand in 1997, receiving a paltry $300 million for it.[4]

These examples show the difference that difference makes. McDonald's realized that the differences between its product lineup and those of its competitors were disappearing. The value-pricing fiasco forced it to think again about where the difference lay, and it wasn't in the food, the pricing, or the storefronts. Quaker diluted the value of the Snapple brand when it watered down its taste and advertising. Perhaps even by distributing it broadly, Quaker made Snapple appear less special and thus less valuable.

Types of Asymmetry

The baseball analogy helps us understand the two kinds of differences, or asymmetries, there are. Through its strict rules, baseball sharpens the contrast in talent and capabilities among individual players. Of course, there are differences among teams, but they lie mostly in the experience, talent, and capability of the individual players and, to a lesser extent, in how well they work together as a team. The differences among teams are (mostly) inherent in the individuals. These inherent differences I call *natural asymmetries.* Those few differences that a coach or team manager might try to develop through training are *created asymmetries*—the development of particularly close teamwork among pitcher, catcher, and third baseman, for instance.

The search for competitive advantage in business is the search for natural asymmetries or created asymmetries (see Figure 4–1). These asymmetries provide the opportunity for competitive advantage. When Quaker reduced Snapple's *difference,* it eliminated the opportunity to create competitive advantage.

Natural Asymmetries

Natural—or discovered—asymmetries are differences companies can exploit that are like the differences among players on a baseball team. You can't take credit for them; you can only take credit for recognizing them and putting them to use. The classic examples are asymmetries based on geography. Red Lion Hotels (now merged with the Doubletree chain) and Shilo Inns in the West and Pacific Northwest exemplify how the natural asymmetry of geographic location can lead to the differentiation of something as commodity-like as a midpriced hotel room. Alaska Airlines' service of the Alaska–Pacific Northwest and the California north-south corridor also takes advantage of a natural asymmetry: the north-south line of highly populated cities that lie

Figure 4-1. Natural versus created asymmetries.

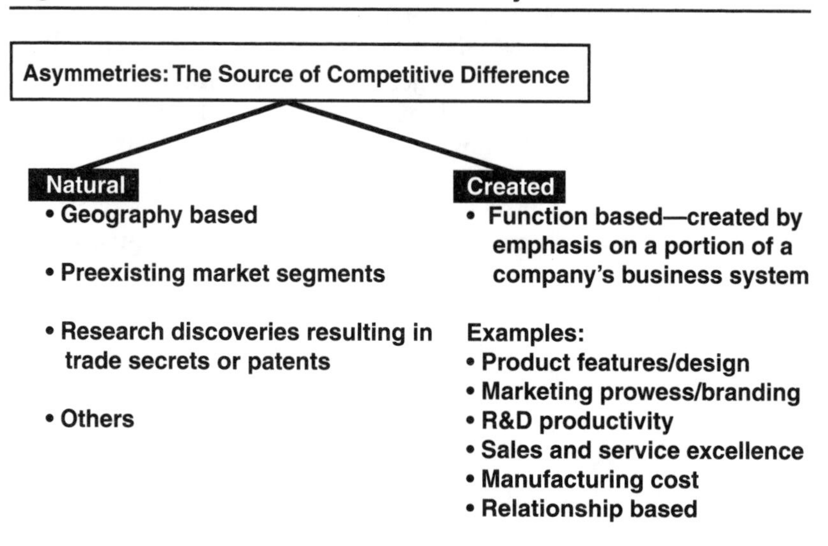

along the Pacific coast from Anchorage to San Diego. It is natural for Alaska Airlines to push its franchise further along this line of longitude to Cabo San Lucas and Puerto Vallarta, along Mexico's Pacific coast. Trucking companies that specialize in specific corridors such as Chicago–Indianapolis, Denver–Salt Lake City, or Munich–Hamburg take advantage of the natural asymmetry afforded by the proximity of the paired centers and the developed highway networks linking them.

Geographical asymmetries are classic, but there are other natural asymmetries where you might least expect them. Research conducted almost half a century ago by GEICO auto insurance revealed that government employees in the Washington, D.C., area had far fewer accidents than other drivers who drove the same number of miles. GEICO began peddling insurance policies at favorable premiums aimed at this segment. It developed such a lock on the segment that it soon attracted the attention of legendary investor Warren Buffet, and GEICO became part of Buffet's Hathaway empire. I call this a natural asymmetry because the risk differences between government employees and the general population were there, waiting to be discovered. GEICO did nothing to create them; it did everything to exploit them.

Likewise, patents based on the discovery of new chemistries provide pharmaceutical companies with natural asymmetries to exploit or ignore. A mining company that discovers a lode of ore in New Guinea and, through negotiations with the local government and tribes, obtains the exclusive rights to blast the mountainsides for the next fifteen years is in the same category.

It's easy to see how a strong company can be created if you're discovering or exploiting a natural asymmetry. As in baseball, you're bound to be a winner when you've got a concentration of natural talent on your side.

Created Asymmetries

The picture is completely different for created asymmetries. If you can imagine a change in the baseball rules that would allow you to concentrate a number of players in right field or vary the batting lineup at will, you can appreciate the impact such a created asymmetry would have on baseball as a strategic game.

Created asymmetries come about because an entrepreneur or a company concentrates resources somewhere in the business system chain. You can find or create an asymmetry anywhere along the business system—from R&D and raw materials purchasing through after-sales service. Figure 4–2 provides some examples.

The Attraction and Disillusionment of Product-Based Asymmetrics

When the question "What makes us different?" is asked in marketing departments and boardrooms, the first responses that come to mind relate to product features and cost. I've presented the concept of asymmetries to show that there are many chances to create difference *outside* these important but overrated areas.

The strengths of product-based asymmetries are obvious. For the same price, who in her right mind wouldn't prefer more for less? But everyone knows that product superiority is neither necessary nor sufficient for competitive advantage. A well-known cell phone manufacturer I've worked with is faced with the frustration of having a superior product—longer battery life, smaller size, even voice recognition—but still being unable to gain more than a small sliver of the market pie. The reason? The concentration of effort on product features was not vital for this market. The company needed, instead, to concentrate on developing *carrier* (telecommunications company) partners to expand its range of service. And it should have paid attention to its brand identity: In one part of the world, its brand name was stigmatized as a synonym for *cheap* and *easily broken*. And this is a market where research has found that a common initial motivation for cell phone purchase is the desire for security.

Product features are a slender reed to build competitive advantage on, even when it comes to health. A company I once worked with

Figure 4-2. Leading companies provide examples of asymmetry creation.

	R&D	*Product Design*	*Manufacturing/ Operations*	*Marketing*	*Sales*	*Service*
Company example	*RadGen (disguised)*	*Sony*	*Nucor*	*Procter & Gamble*	*Nordstrom*	*Les Schwab*
Created asymmetry	Influencing industry standards to favor expected new patents	Clever product design; unwavering attention to Sony brand image	Cost efficiency in steel through the use of scrap, high technology, nonunion labor, and performance-based pay	Excellence in brand management, consumer marketing, and managing promotional campaigns (*Example:* forcing advertising agencies toward success-based billings)	Focus on store environment and sales clerk training and incentives	Local tire chain, charges premium prices for a commodity product by providing fast, consistent, no-questions-asked service no matter what tire brand is sold at numerous locations

invented an at-home diabetes test. What a remarkable instrument that was! For the first time, it allowed people with diabetes to tell whether their diet, exercise, and medication regimens were working. Yet sales to consumers were agonizingly slow. Reason? The company was experiencing the familiar S curve of product adoption. Only *early adopters* would try the product at the start. The rest of the market was waiting to see how these people fared, waiting to see what their doctors would say, perhaps even doubting that the test would perform as promised. But an untold number of diabetes-related amputations, losses of sight, and seizures might have been prevented, or at least postponed indefinitely, if the test had been more widely adopted.

Moral: As a rule—and there are exceptions—don't build your strategic house on the swaying reed of product superiority. This is an old moral, but one that must be relearned repeatedly.

And there is a new reason for it to be relearned. The fungibility of technology and knowledge today ensures that no new product is without competitors for more than a few weeks or months. Consider 3Com's Palm Pilot series of electronic personal information management systems (PIMS). In the late 1990s, it took the market for PIMS by storm due to its nifty form factor and unique operating system. But by 1999, there were nearly a dozen competitors from Philips, Casio, Compaq, Sharp, and others. Prices ranged from $199 to $649, depending on brand and features. Even intellectual property rights such as patents seem to be no great barrier to determined copycats.

Among consumer electronics companies, Sony is the outstanding success story, where product design and features are paramount. And Sony often succeeds not by having the best feature set but by being first in category: Walkman, Watchman, Betamax, Mavicam digital camera, and the HiFD (high-capacity floppy disk) are among Sony's market-commanding firsts.

Cost Asymmetries: You'll Need Wax in Your Ears

The siren song of cost is the second most commonly thought of created asymmetry. How sweet the melody! If we produce the same thing for a lower cost, we'll have an advantage over the competition. The converse is equally appealing: If we don't have costs that match our competition's, we'll be in hot water. But if we build for scale, we'll get immediate lower costs and then secondary lower costs as the learning-curve effect drops in. This is a time-honored concept.

Boston Consulting Group was founded on penetrating analyses of how cost and its stepfather, scale, drive industrial behavior, industry structure, and even economic cycles. Even today, companies genuflect at the altar of low cost.

But although cost will forever be important, cost is changing. In the new economy, cost is mutating. First, in very large projects, suppliers, customers, and even competitors must cooperate. Boeing can't afford to research, design, and build a new jetliner without up-front commitments from the large airlines and coinvestment from engine and avionics manufacturers. So it gets hard to say whose costs have to be low, to determine what the costs are, and to differentiate between *costs* and *investments*.

Second, more and more often, prices are subsidized. In the cell phone business, for example, it's common for the carrier (the phone company) to pay the cell phone manufacturer a subsidy so that the retail price of the phone is low enough to be enticing to consumers. The marginal value of the cell phone service is so high, and the marginal cost to the carrier so low, that it pays to put an essentially free pay phone *hardwired* into the carrier's network service into as many of our pockets as possible.

The advent of the free or almost-free personal computer, with the supplier relying on an Internet connection fee contract or the sale of mandatory advertising-viewing eyes, converts what would ordinarily be a consumer purchase into a distributor's investment. So what is the cost of the PC? Does it matter? To what extent? If the personal computer actually survives the information-appliance revolution, having a low-cost PC may be only the ticket to entry into the new game—a game in which the PC is merely the terminal for other revenue-producing opportunities such as mail-order sales, auto insurance, and utility bill presentment.

Third, another peril of staking your company to a low-cost strategy is that you may get what you asked for. Even today, the Internet allows a consumer or an industrial buyer to compare prices quickly. On a commodity item such as coal, books, a mortgage, or a roll of steel adequately specified, buyers can search for the best price easily. Auctions are now being held in which buyers post requests, and at a set cyber-signal, qualified manufacturers have the privilege of bidding against one another on fulfillment of those requests, down to a preset deadline.

This means that in the future, the best—read *lowest*—price will be the *only* price for a commodity. You might therefore think that if you were the lowest-cost producer and could make a profit at that low price, you would be sitting pretty.

Sometimes. But more likely, there will be somebody out there who offers what seems to be an irrationally low price—a price below your expectations and *their* cost. Reason? There's always somebody liquidating an overstock; always a country with an undervalued currency; always a broker who got the item your customers are seeking at a nickel on the dollar and is willing to sell it for fifty cents. Only now, the Internet makes the world these folks' market. Can a strategy based on cost asymmetry compete here? The warning flags are up.

Fourth, I am one of those who believe that markets are fragmenting. People are beginning to want factory-perfect—but customized—everything. And they are just beginning to get it. Today, professors can order textbooks customized to include just the chapters and articles they want. Music lovers can order CDs with just the tunes they want on them, so long as the artist isn't in contractual slavery to a major label. And the chinks are showing in the major labels' armor, too, as they scramble for industry standards to protect their copyrights on digitally downloaded music.

Toyota has been talking about short runs of a dizzying number of model types, so that buyers will only rarely run into clones of the models they bought. One cell phone maker has cleverly designed a common platform for many of its models, so that various models can be customized via software commands. Newspapers used to be one size fits all; the maximum customization involved *zoned* editions, with modest differences in editorial content and a little more than modest differences in advertising. But now, anyone can sign on to MSNBC or Yahoo! and get the news tailored to his or her interests.

Mutual funds are exploding in variety, if not in performance. And new machines can produce one-of-a-kind plastic prototypes of practically anything, based directly on computer-aided design (CAD) inputs. It's not hard to see that soon a consumer will be able to order a coffeepot or a car seat exactly to desired dimensions. (The resale value of that forty-eight-gallon coffeepot is, I guess, another matter.)

Forever, your salespeople will tell you that the main reason they can't sell more is cost. Forever, the numbers from manufacturing and operations will tell you that more volume means lower cost. But that world is becoming smaller and smaller.

Mental Asymmetry

When cost and product fail to make a difference, many firms—especially service firms—retreat to a created asymmetry called *relationships*. Law firms, consulting firms, even retailers and distributors, all want to build lifelong relationships with their customers. This difference, they hope, will see them through. I wish them luck.

First, it's important to realize that relationships in the business world are just a kind of *branding*. Both are built on the fundamental human need for trust when the stakes are high. I pay more for the branded aspirin than the house brand because (I think) I can trust Bayer over generic America's best. I go to the biggest law firm in town (typically) because I can't give them a tryout in defending the suit against me first; I have to find somebody I can trust to do the job right the first time. If I've had a past successful relationship with Wynem and Dynem, that is evidence I can trust them, and I'll pay them a little higher hourly rate than I would pay to somebody else.

What we're learning in the new economy, however, is that branding in general and trust in particular are becoming asymmetries that can't be relied on. In the gentlemanly world of management consulting—where I've spent much of the last two decades—*shoot-outs* and *beauty contests* among competing firms are becoming much more common. Companies are becoming sophisticated. Many have had ten, twenty, thirty years of experience with consultants and have a few former hired guns on board.

Second, cheap transportation and cheap, fast telecommunications make it possible for professional service companies to compete for and complete assignments globally. Recently, I worked with a Boston-based strategic market research company on an assignment for a Seoul-based conglomerate interested in telecommunications markets worldwide. We were last-minute party-crashers to the consulting-contract bidding process. Cheap transportation, a nonexclusive agent in Korea, and Internet-based telecommunications—and, I must say, an outstanding proposal—won us a contract where two years earlier we would have had no chance. By showing our faces on a few days' notice in Seoul, we earned enough trust points to shoulder aside one of the best-known U.S. product-development consulting firms.

Third, it's becoming common knowledge what it takes to become known. As of this writing, the price for becoming well known in the United States is a minimum $30 million to $50 million for consumer goods to start, and anywhere from $10 million to $40 million to sustain (that doesn't count the cost of *overcoming* a competitor's recognition factor).

In 1997, Internet start-ups spent $15.6 million on advertising and public relations; by 1998, this figure had nearly tripled to $44.6 million.[5] From a sample of 300 to 1,000 respondents, market research techniques can now define with high precision (depending on the product and its permutations) what a product or service should contain, the value of various brand names for the product, and how it should be priced for either profit or penetration—to the penny.[6]

This means that there is now a price on that sovereign mental space reserved for brand names once thought too entrenched to be dislodged, such as Coke and McDonald's. They still have formidable brand names: People *trust* them. But is it so hard to imagine Coke or McDonald's one day becoming *uncool*? With enough pounding from other brands or a change in the fashion winds, it seems likely to me that their brands could go the way of another impregnable brand, Cadillac.

Coming Down from the Mountain: The End of Sustainable Competitive Advantage

These disturbing—but I hope exciting—reflections are sufficient to cause the first shivers of doubt among those who believe that corporate strategy consists in finding a sustainable competitive advantage. The SCA has been the holy grail for at least a generation of strategic thinkers, if not two. I do not say that it's dead, but it's certainly confined to special corners of the industrial planet, as are other endangered species. To understand this, it's worth going back to fundamentals for a moment.

SCA is the powerful idea that a company's strategy should be based on those aspects of its business at which it is better than anybody else can be. Usually this means matching a company's assets (broadly defined to include people, skills, intellectual property, brand name, and so forth) to a specific market segment (the famous *niche*), then erecting high *barriers to entry* to guard that market niche. People don't like to use the term, but the idea is that this will result in near-*monopoly* pricing within the market niche. The only limit on prices is that you must make sure you don't raise them so high that it pays quiescent competitors to build scaling ladders to climb over those entry barriers and invade your niche.

A competitive advantage is therefore an advantage related to serving some particular market. The company's assets provide a way of serving the market that's different (in a way favorable to profitability) from what competitors can offer. A large part of strategic work rightly consists in finding a market niche (or *segment*) where your assets match customer needs better than anybody else's, then determining ways to make that match as impregnable as possible. That's how advantage gets to be *sustainable*.

For example, if you build an efficient private transportation network along the corridor from Marseilles to Madrid especially for your very heavy, bulky product—say, granite and marble slabs—you create a barrier to entry for those who would like to ship a competing product along the same route. It's sustainable if you have a way to keep the bulk of your competitors from finding an equally low-cost transportation solution on that corridor. Or, less classically, if you are a software maker with an installed base of users, you can reinforce your users' reluctance to install your competitor's software by making your applications smoothly compatible with one another but much less compatible with those of the competition. Or, like RadGen, you can try to get the international standards committee to adopt rules that favor the steps you've taken in technology. Once the rules are in place, you can work to keep them rooted, through legislation, lobbying, and ensuring that the bulk of the customers use only compatible technology.

In every case, the barrier you've erected and are trying to sustain is based on a difference that matters to the market segment you are serving. In the preceding examples, there are differences in transportation costs, differences in systemwide usability, and differences in the ability to match or exploit a technology standard, compared with those of the competitors.

These differences are either discovered (natural) asymmetries or created ones. Yet neither category (except in some special cases) holds much promise for providing *sustainability* in the coming century. Advantage in the new century will be fleeting. If sustainability crumbles, so does the strategic edifice called SCA. And this means that SCA should no longer be the touchstone for the quality of strategic thinking.

My thesis is simple: Many of the differences used in the past to erect sustainable advantages are disappearing due to technology, transportation, communication, and the globalization of culture. A few remain for a few companies and industries—the lucky few survivors of the endangered species. The examples I'm going to cite now are exciting, even though they are familiar.

Geography, once a source of natural asymmetry for many industries, is becoming irrelevant. Obviously, transportation of physical products is cheaper than ever before. Seafood for restaurants is nearly as fresh in San Antonio as it is in San Francisco. Denver manufacturers and agribusinesses now compete in worldwide markets, even though they are landlocked in the center of the continent. Car parts are made wherever manufacturing them is cheapest, then transported and assembled where they are needed. The magnitude of this change is hard to exaggerate: Over the past two decades, the cost of transporting goods from Asia to the United States by ship has dropped by *two-thirds* in inflation-adjusted terms. That makes seaborne transportation about 1 percent of the value of consumer goods. As ships get even larger—perhaps twice the size of today's largest—unit costs will drop even further.[7] Mail-order houses are now competitors for department stores, due to cheap air express. Federated Store's Bloomingdale's unit has growing catalog sales to Japan—fulfilled in Connecticut. People are transported worldwide, too, and as national barriers fall, talent and technical know-how are globally marketable resources.

Less obviously, cultural differences, a subset of geographical barriers, are making heretofore sustainable geographic barriers shaky. The English language and American culture are the lingua franca and cultura franca. This makes it possible for descriptions of dresses and blouses in Bloomingdale's Japanese catalog to be printed in English. Even the order forms in the original editions were in English. This means that Japanese retailers have a new—and perhaps unexpected—competitor for the housewife's or the young professional's disposable yen.

At a recent meeting in Seoul, the conglomerate manager I was assisting asked his secretary to order in lunch. Twenty box lunches arrived on the marble table in short order. The Americans ate Korean box lunches from a take-out shop nearby; the Koreans voted for Burger King in a landslide. The local sandwich emporium has an overseas competitor. We can imagine the chagrin of the mom-and-pop entrepreneurs who for twenty years have exploited their advantageous location proximate to the headquarters of this famous *chaebol* (family of closely related industrial companies). They have successfully defended against a Korean competitor five blocks away with long days and nights of hard labor, only to see their business chopped in half by an overseas brand.

Communication is cheap, fast, and reliable. Insurance companies process American claims in Ireland using well-educated but low-cost labor. The transfer of funds across time zones is remarkably easy. Being close to a home market is no longer a source of advantage, nor is ready access to capital markets when two companies of similar creditworthiness are being evaluated.

Talent is now mobile nationally, continentally, and globally. I have an ethnic Russian friend, an investment banker who learned his trade at Salomon Brothers and Morgan Stanley. He is now ensconced in Moscow with Deutsche Bank. When GM decided to build a low-cost factory in Argentina, it pulled talent not only from South America but also from Europe and North America, as well as from competitors such as Toyota. A Denver start-up I know flies in half its executive officers every week from Minneapolis, Los Angeles, Tampa, San Francisco, and Portland, Oregon.

Intellectual property rights, such as patents and copyrights, no longer confer a long-term competitive advantage. Threats range from outright piracy (as practiced with software in China) to the ability to circumvent patents to the abysmal state of U.S. patent law and the protracted legal process that can bankrupt a patent holder. In November 1990, a court ordered Ford to pay $10.2 million and Chrysler to pay $11.2 million in royalties to a solo inventor for the intermittent windshield wiper system that he had invented and Ford had used, but the case took near-fanatic patience and pursuit and years to resolve.[8]

A client of mine spent thousands of dollars on a license to a key patent based on university research. When the time came to sell the company, its CEO estimated the license to be half the company's total value. A patent attorney had researched it, the Patent Office had approved it, and the whole industry respected it. Yet a half hour's research by a prospective buyer proved the patent to be invalid, and thus the license to be worthless. The company was liquidated, not sold. A patent, with a lifetime of seventeen years, is less and less often a source of sustainable advantage.

Relatedly, superior technology is not only moving fast but is also so fecund that what Michael Porter calls *substitutes* for a company's products arise very quickly. Even if a technology is proprietary, it is quickly obsolete or irrelevant.

Scale economies, which once protected large companies, are crumbling, industry by industry. An example of this is the computer technology for book printing that allows economical runs of only a few hundred or a few thousand books. A few years ago, this technology was an industry miracle. Yet in 1999, a new company started up that publishes *single copies* of a standard 9- by 5-inch paperback with full-color cover and sells them economically to consumers for under twelve dollars. This newer technology means that in the future, publishers can dispense with nagging economical print runs of 2,500 or 5,000 or 10,000, and thus with the traditional services of printers. Linked to the Internet and express delivery services, they may soon dispense with physical distribution services (pickup, warehousing, and distribution) of such book distribution giants as Ingram and Baker & Taylor in the United States. College texts written for a broad market are becoming obsolete, as professors can now order custom texts (featuring, of course, their own scholarly publications) in quantities matching their class sizes.

New machines are being developed that can make one-of-a-kind examples of increasingly complex products, and as these machines become cheaper and the associated design software becomes easier to use, I foresee the not-too-distant day when large factories are practical in only a few industries and the mom-and-pop neighborhood craftsperson returns.

This affects a hardware manufacturing operation I own. My customers must now buy our injection-molded products in lots of 2,500; we must make them in lots of no fewer than 25,000. But soon they may be economically producible in lots of a hundred or ten or even *one.* This means that the cost of our injection-molding tooling will no longer be a barrier to others—even our own customers—entering the business.

How does all this shake the foundations of SCA? Suppose the *givens* for your proposed strategy depend on low-cost production in a mass-production environment. If a competitor gets wind of your plans early enough, it can set up its own low-cost factory; it's likely that raw materials can be shipped cheaply, experts from competitors or universities hired globally, and the globe scoured for the cheapest capital from London, Hong Kong, New York, or the United Arab Emirates. A new factory can be set up either close to a market or where labor costs are most favorable, and it can use cheap transportation to ship finished goods globally.

This is not to say that in the end you might not have the lowest-cost production. But it does mean that at the very least, the margin by which you are lowest cost is thinner than it's ever been. It also means that although you might be lowest cost for now, it might not be long before your production costs are matched by a firm you've never heard of in a country whose name you can't spell, or by someone down the block.

Finally, the concept of sustainable advantage is under threat from the fundamental human drive to be creative. The difference in today's world is that there is such a diversity of basic science, applied technology, research data, communications technology, and new materials to be creative with. The more information and tools we have to be creative with, the greater the opportunity for creation. The easier and cheaper global and international communication is, and the greater the ubiquity of English, the more opportunity there is for long-distance and international collaboration in research, development, invention, and design.

In short, little can withstand inexpensive global transportation, a single world language, cheap communication, a free flow of talent, paltry intellectual property protection, and the decline of scale economies. In classical economic jargon: The opportunity for substitutes for your product or service will intensify. In classical boardroom jargon: Where the devil did *that* come from?

The New World Order

In 1999, the Japanese economy was down in the dumps, with flat growth and an official unemployment rate of 4.8 percent. The *real* unemployment rate, not to mention the *under*employment rate, was much higher. However, to take the long view, Japan is very much the success story of the postwar period. Don't count the Japanese out for long.

Remember, Japan is net owner of foreign assets, whereas the United States owns fewer assets abroad than are owned by foreigners in the United States.

I put the current Japanese recession in perspective because Japan's economic growth is striking evidence for the thesis that the venerable *sustainable competitive advantage* has entered its dotage. Japan is a country with almost no natural resources. It must import over 99 percent of its mineral ores, 96 percent of its mineral fuels, 85 percent of its wheat, 70 percent of its corn, and 80 percent of its barley. In foodstuffs, it is self-sufficient in rice, due to the political clout of its farmers. In 1988, it imported 93 percent of the nickel ore available for export on the world market, 57 percent of the copper, and 30 percent of the coal and iron. Japan imports 5 tons per capita to the United States's 2 tons. To keep Japan supplied, every merchant vessel in the world has to devote at least one full trip per year to feed Japan's need for raw materials.[9]

But consider this. The Japanese export 1 pound of goods for every 8 pounds they import. Yet they are able to add so much value through brains, courage, sacrifice of living standards, and hard work alone, that that 1 pound has more value on the world market than the 8 pounds imported: Japan runs a current account trade surplus.

This goes to show that even that stalwart mainstay of SCA, gaining an impregnable cost advantage through privileged, low-cost access to raw materials, means very little in the new world economic order.

This is my hypothesis about SCA. Falsify it if you like; falsify it if you *can.*

Where SCA Survives

SCA will never be extinct. The way things are headed, it will, I predict, survive in scattered places. Lucky is the company (or country) that can latch on to one of the dwindling differences that can provide a foundation for SCA. Here are some of the possible survivors that are not to be overlooked:

- *Differences mandated and protected by the force of government or force of arms.* With the advent of the North American Free Trade Agreement (NAFTA) and the European Union and the Euro, trade barriers, a species of difference, are dwindling. But others survive inside borders. Governmental programs, subsidies, and set-asides can be sources of long-term difference. (But woe unto those who fail to reckon with the capriciousness of representative or single-party government.)
- *Differences mandated by nature, such as low-cost, privileged access to minerals.* Although I believe that this difference is shrinking for con-

sumers of minerals (as argued by the example of Japan), there is no doubt that mineral producers with privileged access will have a continued advantage. Usually, these privileges are abetted by force of government. For example, large oil companies must gain the favor of governments of developing countries and emerging remnants of the Russian empire, such as Azerbaijan and Kazakhstan, to explore and pump oil from their territory. Those who haven't made the requisite political, personal, and monetary arrangements are likely to be cut out of the action for decades.

- *Intellectual property rights.* Occasionally, larger companies will be able to carve out some sustainable advantage through patent portfolios and battalions of lawyers deployed globally. As Jacqueline Susann wrote, however, once is not enough.

- *Increasing returns.* The business world is now testing the hypothesis that the phenomenon of increasing returns made prominent by Internet commerce will be a source of SCA. This is the idea that once a product or service (especially information technology) becomes used by a critical mass—a plurality or a majority of users—it tends to dominate the marketplace. It then provides monopoly or monopoly-like profits for its owners. The classic examples are Microsoft's operating systems and office suite software. It is just far easier for those who don't have these systems to convert to them when they must communicate (be *compatible*) with those who do. Another example is the global use of the English language, whose remarkable market share provokes even more widespread adoption.

Strategy books such as *Net Gain* promote this last theory.[10] I believe that in the short and perhaps the medium term, it will hold true for those limited products and services for which interchangeability and compatibility are paramount. However, I think the application of this concept is limited.

Nearly every product has to match some set of industry standards, and this limits variation. For example, European cabinetry is generally standardized on a 32-millimeter grid. This means that shelving, doors, drawers, and uprights are cut and drilled in 32-millimeter increments. Thus, cabinet components can be connected to one another with hardware of standard configurations and are (generally) mutually compatible. This doesn't, however, prevent the existence of hundreds of cabinet manufacturers and dozens of hardware suppliers and woodworking machinery manufacturers, all competing fiercely with one another.

In the Microsoft case, it is the ownership of the standard and the monopoly on access to the standard that has reduced its competition.

But I believe that this will not last much longer. If the U.S. Government doesn't bust this monopoly on antitrust grounds, the market will. First, the all-purpose desk- or lap-bound personal computer will not be a thing of beauty in consumers' eyes forever. Even now we see the emergence of personal digital assistants (PDAs) like 3Com's Palm Pilot and Psion's pocket computer, using their own operating systems yet synchronizing data with Microsoft-powered machines. Second, restless creativity will make Java or something like it the Esperanto of information technology, allowing people to stick with Microsoft if they like, but giving them the option. Third, future telecommunications standards will reside not within the machines that individuals own but in the networks that connect users.

Microsoft is actually a unique case of *owning* the standard that everyone must purchase a copy of, rather than merely adhere to, and it's a standard that has almost zero variable cost (including customer service), making it possible for Microsoft to sell at very low costs to gain market share. Also, there's strong evidence that even the software industry's 800-pound gorilla, Microsoft, can't monopolize a market without highly competitive products. Stan Liebowitz and Stephen Margolis examined fifteen years of software product reviews. They concluded that Microsoft wins high market share when the reviews' consensus is that its products are better. It loses when they're inferior. Two examples are Microsoft Money versus Intuit's Quicken in personal financial software and the Microsoft Network versus America Online. Similarly, the two professors also debunk the business school shibboleth that Matsushita trounced the superior Sony Beta product in VCR formats through sheer marketing prowess. Matsushita's VHS format had a key superiority favored by coach potatoes: longer tape recording time.[11]

For these reasons, although I see the phenomenon of increasing returns as a viable source of sustained advantage in some circumstances, I see it as a limited one.

And the same is true of SCA as a whole. To be sure, I have nothing against finding or creating a sustainable advantage when you can. If you see it, do it. You will be one of the lucky few. But the main message is that you mustn't be fooled by the appearance of sustainability based on a hopelessly optimistic, quietly obsolete view of what barriers are sustainable. (See Figure 4-3 for a schematic view of why SCA is under pressure.) Better, I suggest that you cast your lot with *nimbleness.* What kind of nimbleness? That's the topic of the next chapter.

A Home Inspection Checklist

It's usually wise—if not required by law—when buying a home to have a professional inspect its foundation, wiring, plumbing, and roof. So it

Figure 4-3. Where sustainable competitive advantage survives, it is still under pressure.

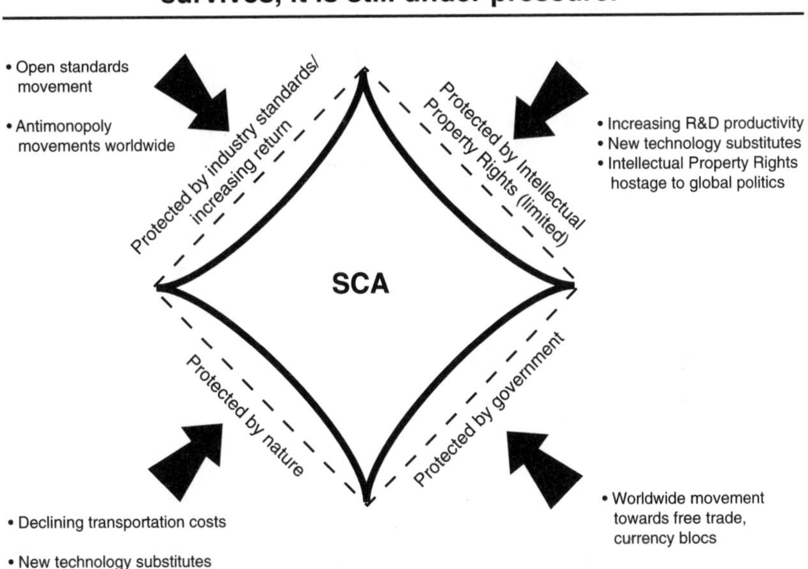

should be with strategy, to determine whether there is a sustainable difference for your current or proposed strategy to rest on. Here's a checklist that I suggest your strategy team get busy with tomorrow.

1. The three main characteristics of our current market niche that make us really superb (or even moderately competent) competitors are: (RadGen's answers are given first, as a sample, followed by blanks for you to fill in.)
   a. A small number of large customers.
   b. Customers who appreciate technology but also appreciate service.
   c. A market (the video display market) that is expanding in double digits each year.
   a. ______________________________
   b. ______________________________
   c. ______________________________
2. The differences between us and our competitors that make us one of the best, if not *the* best are:
   a. We've been in the market since the industry's start; people know us, our technology, our brand, and our reputation.
   b. Our technology has been superior, at least until recently.
   c. We're huge; we have scale in manufacturing, sales and service, and R&D.

a. ______________________________
b. ______________________________
c. ______________________________

3. Are these differences natural or created?
   a. Created, but we've been here so long, it seems like our being the leader is a natural part of the industry's landscape.
   b. Definitely created; we've worked hard and invested aggressively in manufacturing and R&D.
   c. Created, but again, once our plants and people are in place, they're almost like natural asymmetries we can depend on; after all, our plant and equipment are almost amortized.

   a. ______________________________
   b. ______________________________
   c. ______________________________
4. For the differences that are natural, how would a competitor overcome them? For the differences that are created, how would a competitor attack them?
   a. A public relations campaign by competitors and an infusion of capital by a venture capitalist or acquirer would give number 2 and number 3 a lot of credibility and staying power, and capital is cheap right now.
   b. Equivalent or superior technology by competitors is making customers question our leadership; it seems as if we have to prove our leadership every day, like the front-runner in a political campaign—the guy everyone attacks.
   c. Actually, we don't know whether we're vulnerable on the cost side. We *think* our manufacturing costs per unit are lower—they should be—but we haven't really looked at that. Then again, perhaps number 3 and number 5 are sourcing some manufacturing steps from overseas. Our sales and service costs per unit are probably about the same, because we spend more in absolute dollars on sales, even though our volume is higher. That's how we keep our advantage.

   a. ______________________________
   b. ______________________________
   c. ______________________________
5. All in all, can these asymmetries be sustained?

   RadGen says: From this analysis, it looks as though we're running faster and faster to stay in the same place. We have to overinvest (compared with competitors) in R&D just to keep ahead, because of our history. On the sales side, it's the same story. Our advantage in service has been bought by investing more per unit sold in sales and service, but that does seem to secure loyalty. We'd better test how much loyalty we're getting from that extra investment—or are we just buying our market share?

You say:

____

____

____

Notes

1. *Business Week,* 17 March 1997, 38.
2. *American Demographics,* July 1997, 64.
3. *Brandweek,* 31 March 1997, 54.
4. "The Honeymoon's Over," *Sales and Marketing Management,* June 1997, 54–61.
5. *Advertising Age,* 25 January 1999, 3.
6. George Boomer, "Conjoint Analysis vs. Discrete Choice Modeling" (Newton, MA: Strategy Analytics, 1998).
7. Stewart Taggart, "The 20-Ton Packet," *Wired,* October 1999, 250–54.
8. Joseph P. White, "Chrysler Must Pay Royalty to Pioneer of Wiper System," *Wall Street Journal,* 12 June 1992, A3.
9. George Friedman and Meredith Lebard, *The Coming War with Japan* (New York: St. Martin's Press, 1991), 163–65, 167–68.
10. John Hagel and John Armstrong, *Net Gain* (Boston: Harvard Business School Press, 1988).
11. Lee Gomes, "The Truth about Marketplace Battles," *Wall Street Journal,* 26 August 1999, A16; Stan J. Liebowitz and Stephen E. Margolis, *Winners, Losers and Microsoft* (Oakland, CA: Independent Institute, 1999).

Chapter Five

Making Strategy Dynamic: The Dawn of Opportunity Creation and Exploitation

Everything flows.
—HERACLITUS

In chapter 4, I pressed the case that seeking a strategy based on *sustainable competitive advantage* is becoming futile and may be dangerous to a company's strategic health. I cited changes that have forced this conclusion. But it would be naïve to think that these changes affect all industries and companies equally. Technology-communication-globalization is not an equal-opportunity destroyer.

This chapter proposes a replacement for SCA. I call it *opportunity creation and exploitation* (OCE). Unlike SCA, which is a proposed ideal state and a happy situation for a company, devoutly to be wished for but seldom achieved, OCE is an organization and mental ideal: a frame of mind for the pursuit of a successful strategic hypothesis and a touchstone for the allocation of resources.

In this chapter, I

- Categorize broadly the variety of strategic environments that exist for industries and companies—with some help from the world of mathematics.
- Describe OCE and the OCE cycle.
- Suggest ways to match the OCE cycle to your needs.
- Use some exciting field research on evolution to hold up a mirror to these concepts.
- Lay out a template for making the OCE cycle a practical part of your management habit.

Variety and Diversity in the Corporate Garden

Many people who devote themselves not to the world of commerce but to the world of science, especially biology, were originally ensnared and delighted by a single facet of that world: the incredible diversity of life-forms. My life in business has alerted me to a similar diversity in the roads that companies can take to success. There is no one right way to do things: Diversity is the rule. For example, consider the spectacular success of the straitlaced Ross Perot companies. EDS, which Perot sold to General Motors in 1984 for $3 billion, and more recently Perot Data Systems, which went public with sales at $780 million,[1] were studies in conformity: how to sell, where to live, how to dress, and even, I'm told, how to date and how not to divorce. The contrast couldn't be greater with the early days of Apple Computer, where chaos and eclecticism—not to mention petty tyranny—were on the daily menu.

No one can argue with the success of either. Corporate-style autocracy can be successful, even though today "collaboration" is cried hourly from the minarets of our business schools and $600-a-day seminars. No one can argue with the success of General Electric, which recruits executives prone to argue their views loudly and in phrases peppered with profanity; yet much has been accomplished by and more is expected of Rick Thoman, CEO of Xerox, who is a master of soft-spoken intensity.[2] And as I suggested in chapter 1, some companies achieve success by "sticking to their knitting," but many others flit from market to market and even from industry to industry. In 1998, Zapata Corporation, in the oil exploration and fish wholesaling businesses, attempted to buy thirty-one Web sites and become a big-time Internet player. Unfortunately, the plan was soon abandoned.[3]

The point is that diversity and variance rule; similarity and conformity fail. But as in biology, it's instructive and useful to find order. The fossil record in evolution and the taxonomy of current species show wonderful layers of order. It is worthwhile to try to impose some broad order on the diversity of business situations here.

Sutherland's Buckets

Fortunately, a robust conceptual framework developed by John W. Sutherland is available to help us. It doesn't merely fit with the concept of strategy as a falsifiable hypothesis but deepens its meaning and usefulness, as well.[4] Recall from chapter 1 that a hypothesis is an *if-then* statement in which the initial phrase sets out the conditions under which we expect a particular result. For instance, in our earlier example from Wal-Mart: *If* we cut logistics costs by 10 percent and pass the savings on in the form of everyday low prices, *then* market share will increase by at least four points, and the resulting volume will improve same-store gross profits 5 percent more than current pricing. Sutherland focuses on the *strength of the link* between the conditions and the likelihood of the resulting event. He sorts links into four categories according to the links' *inherent variability*: the *deterministic*, the *moderately* variable, the *severely* variable, and the *indeterminate* categories.[5]

Deterministic

For a given set of initial conditions (or business actions), there is only one outcome that is highly likely. The classic example is a mechanical system such as a steam engine, where the forces are so bound by strong valves and walls of steel that an initial condition (the conversion of water into steam) makes only one result (a rotating flywheel) highly likely.

In the business world, companies lucky enough to possess a sturdy SCA are most likely to have deterministic situations. Coca-Cola's periodic marketing promotions come close to having deterministic outcomes. Long experience dictates how to craft promotions, where and how much to spend advertising dollars, and, within a relatively narrow range, what the result will be. For example: If we spend $40 million ($25 million on TV spots, $8 million on magazine ads, and the remainder on merchant co-op programs), making sure we feature celebrities appealing to the twelve- to nineteen-year-old age group, then it's highly likely that we'll see a 4 percent bump in sales over the 2001 Labor Day weekend period (numbers are disguised).

In natural-resource industries, economists are usually able to tell at what prices customers will start substituting steel for aluminum, for instance. Such industries are so well studied, the driving forces so well known, the costs so public, and the related behavior so rational that economists can predict accurately, at least over the short term, which bauxite smelters will reopen at what worldwide aluminum price and volume ranges. In my work in the aluminum industry, it was patently clear how the short-term demand for aluminum—primarily for beverage containers, and secondarily for aircraft and autos—would affect the price. And from that, it's an easy step to predict probable competitor behavior. Since the cost and capacity rank of each of the world's aluminum mills are well known, it's easy to predict what new supplies will come on stream over the next few years and the range of the corresponding effect on prices.

I don't want to exaggerate the degree of this determinism. It doesn't have the certainty of death and taxes. But absent war and international recession, a deterministic environment is on the same order of predictability as a public school district's estimate of the number of current ninth-graders who will show up for tenth-grade classes the following fall. (As an aside, this is the sort of determinism that the central planners of the socialist countries attempted to apply to all industries, markets, and behavior—regardless of their inherent variability. Anything less determined just wasn't *scientific*.)

Moderately Variable

For any set of initial conditions, there is a range of outcomes, but the highly likely outcomes are *qualitatively* similar. Another way of saying this is that there is inherent variability in the causal link between conditions and results—between *ifs* and *thens*.

Automobile marketing practices are familiar examples of moderately variable cause-and-effect relations. Since the 1980s, U.S. manufacturers have indulged in periodic rebates, incentives, and price promotions to keep demand up, factories humming, and market share high. Typically, these practices raise volumes, but the results on market share and profitability are far from predictable. Competitors may or may not follow their lead. Consumers, burdened by consumer debt and swings in confidence about their jobs or the economy, respond erratically. But history suggests that consumers' responses, however variable, tend to remain within a predictable range. In other words, on the one hand, it's highly unlikely that a rebate program will have zero or negative effect, but on the other hand, it's highly unlikely that a Chevrolet rebate program will prompt Ford owners to abandon their Mustangs and Tauruses in droves for Camaros and Malibus.

Another example of a moderately variable situation would be the launch of a new antidepressant drug by a large pharmaceutical company. The drug has some advantages, let's say, over its rivals from other pharmaceutical companies. It might require only a single daily dosage, so that patients need to remember only once a day, at breakfast time, to gulp down their medication. It might have fewer common side effects. The pharmaceutical company has launched many improved products before, but not breakthrough ones. Its handsome, well-coifed sales reps can certainly get 20 to 30 percent of psychiatrists to dispense free samples in the first four weeks of launch. It knows that an 80 percent market share in six months for this improved product is unthinkably optimistic (physicians are a conservative lot), but aiming for 10 percent wouldn't be merely pessimistic; it would get the product manager fired.

Severely Variable

For any set of initial conditions, there is a range of outcomes, and they may be qualitatively *dissimilar.* The classic examples here are sports matches. Beginning with an extremely well-defined set of conditions, the outcomes for winner and loser are qualitatively different and, in evenly matched opponents, hard to predict. In business, a strategy based on fundamental R&D or on the launching of a brand-new product into an uncertain market is severely variable. An example might be the risk our fictitious friends at RadGen (see chapter 3) are taking in gambling that their R&D department can make a quantum leap in display technology. If RadGen's senior managers are fully honest, they know that there's a good possibility that Dr. Wing (the head of RadGen's R&D effort) may fail entirely—or succeed beyond all expectations. We are dealing with fundamental research into nature, which is crafty and reluctant to reveal all its secrets. The finest management brains may be able to set the initial conditions carefully, but the results may be wholly unpredictable.

In the United States, a business strategy based on lobbying Congress is also typically severely variable, particularly if a high-visibility issue is involved. The same is true with strategies dependent on judicial or quasi-judicial proceedings. Examples are recent U.S. settlements between tobacco industry companies and the various U.S. states. The tobacco industry has negotiated agreements to curtail marketing of cigarettes in return for indemnification against medically based lawsuits. But this agreement is sure to be tested in court. It's highly unpredictable how narrowly the courts will interpret that indemnity.

Similarly, when private companies accuse others of monopolistic practices by either filing lawsuits or beating their public relations tom-toms, it may seem like a good idea initially. But there's little telling where the effort will wind up. If it eventually gets to court, the plaintiff could win big, lose big, or wind up settling the case. As of this writing, the U.S. Department of Justice's Antitrust Division has Microsoft in the dock on charges of attempting to monopolize the Internet browser software market. All the participants are trying to guess from Federal District Court Judge Thomas Penfield Jackson's facial expressions and laconic comments how he will jump, which is interesting, because even midtrial, he may not know. The range of outcomes is wide indeed—from finding Microsoft guilty and imposing severe penalties to finding it guilty of unfair practices but letting it off the hook entirely, in view of the changing software marketplace.

It's ironic that legislation intended to tidy up relations among people and companies so often creates severely variable situations. In the United States, telecommunications deregulation legislation passed in 1996 was supposed to pave the way for greater competition in long-distance and local telephone service. As of 1999, the country in general was still waiting, along with long-distance colossus AT&T, for the advent of local competition. The Regional Bell Operating Companies (RBOCs—Baby Bells) have managed to stiff-arm AT&T by preventing it from competing with them in local phone markets, even at the expense of their own long-distance ambitions.

The defense industry is rapidly consolidating. Here, government policy has led the industry into a severely variable position. At the height of the Cold War, the Department of Defense (DOD) was procuring twenty different types of combat aircraft. Now this number is down to three. In the early 1990s, the government was concerned that defense assets such as factories and engineering know-how would erode if the paucity of contracts led defense contractors to exit the business. On the projects that remained, R&D was often so expensive that individual contractors couldn't afford to bid on a program and then lose. So the DOD, under then-secretary William J. Perry, let it be known that it would support industry consolidation, even though that meant it would have fewer suppliers—sometimes only one—for key defense systems.

In 1997, however, DOD reversed course and announced that defense mergers would no longer avoid scrutiny as they had in the past. Mergers would now have to be approved on a case-by-case basis. From the point of view of a contractor executive, the situation has evolved from moderately to severely variable. After analyzing and negotiating a potential acquisition, DOD can turn thumbs up or down on a proposal, or even force divestiture of lines of business to potential (or *actual*) competitors.

Deregulation of the once comfortably deterministic world of electric utilities is accelerating the industry's descent into the severely variable. Under deregulation, consumers and industrial users will be allowed to choose their suppliers of electricity. Utilities will have to bid for business, and they will be able to buy electricity wholesale from any willing seller. Mergers and acquisitions—some in the form of aggressive overseas companies such as Scottish Power—will cause some companies to disappear, especially those with a high cost base. But strong political forces will make the environment even more variable than deregulation alone. For example, those who have enjoyed cheap electricity (especially those in the U.S. Northwest, which enjoys cheap hydropower from taxpayer-bought dams) will have to pay more for it. They won't like it, so they will inevitably turn to the political system to remedy perceived injustice and actual discontinuity. And, as we've seen, that's likely to exacerbate the inherent variability of the industry.

Indeterminate

The initial conditions established bear *almost no relationship* to the outcomes. An example might be the fashion industry, where it is very hard to predict which trend will catch fire. Nevertheless, there are those who are nimble enough to make money at guessing what brands, images, and designs will take off from month to month.

For some studios, directors, and producers, the motion picture industry is indeterminate; for others, with the dollars to sign up the few mega-stars, it is merely severely variable. No one can predict what film will be a hit, especially if its credits don't carry some heavy stars as box-office draws. At best, studios can hedge their bets. Increasingly, they bid against one another to attract mega-stars, lavishly advertise and promote new films, and try to negotiate payment guarantees when the pictures go into syndication (to television, to foreign markets, and to video stores). Lately, they have even been using market research firms to tailor their story lines to the public's taste. Nevertheless, motion pictures remain an indeterminate *hit* or *bomb* business.

It's worth identifying the Sutherland bucket in which your company lives—deterministic, moderately variable, severely variable, or indeterminate.

Causality, Predictability, and Inherent Variability

In the following discussion, I talk about the *inherent variability* of various business situations. From the point of view of the strategist operating in a business environment, the degree to which the situation is variable is determined by the unpredictability of outcomes, given specified conditions. And that's what is needed to categorize strategic hypotheses. This is illustrated in Figure 5-1.

Pity the poor strategist. At any given time, the strategist can't know whether a hypothesis isn't panning out because of the inherent variability in the environment, poor specification of the initial conditions (the *givens* and *ifs*) or poor understanding of the outcomes (the *thens*), or both. Further, there is no reason why the variability of a business environment can't change over time. We've already seen some examples of environments that are becoming increasingly variable. Others may be becoming *less* variable. For example, the Internet business environment as a whole may be moving toward less inherent variability as technology and expectations settle down and as winners emerge to begin their (temporary) domination of the Internet landscape.

The only way to extract *inherent* variability from *apparent* variabil-

Figure 5-1. Sutherland's categories sort business/industry environments.

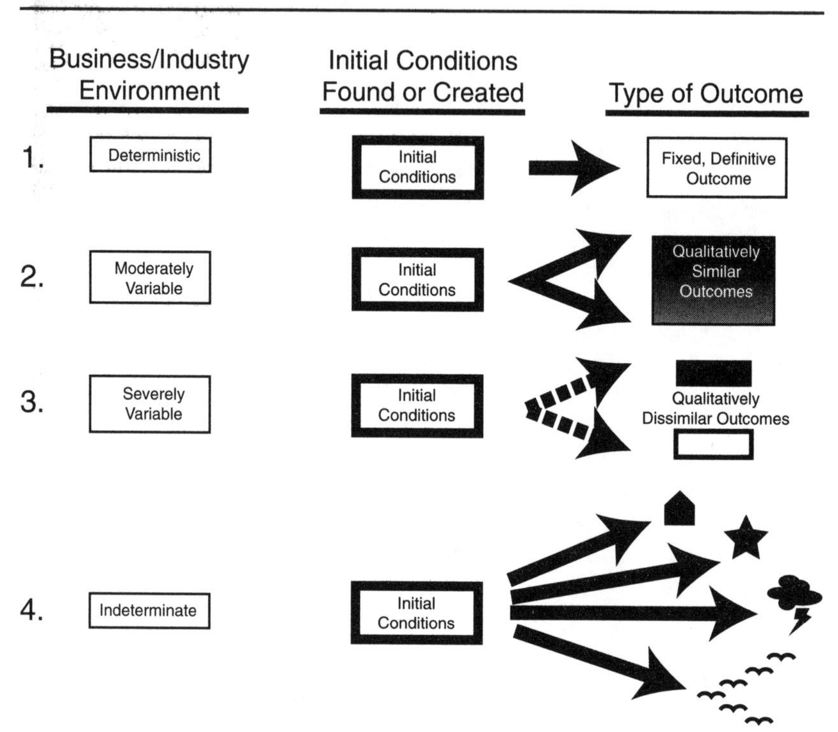

ity is to keep on hypothesizing, to keep on testing. If it does nothing else, persistent testing will, like a sculptor working a rough stone, start defining the shape of what is and isn't predictable—that is, sorting what is inherently variable from what isn't, and thus establishing the boundaries of precision that a strategic hypothesis can have.

In light of the limitations on our knowledge, I refer in what follows to the predictability or unpredictability of business environments—whether merely apparent or truly inherent. So we'll be taking the point of view of a sophisticated, exploratory business strategist, who is nevertheless only darkly privy to whether unpredictability is perceived, inherent, or developing.

Sutherland's buckets are acknowledgments of the fact that there are varying degrees of strength in the linkage between the *if* part of a strategic hypothesis and the expected *then* portion, depending on the degree to which the environment is variable. The more tentative the linkage—that is, the more variable the environment—the less likely it is that a company can build a sustainable competitive advantage.

The reason is that *sustainability requires reliability in the cause-and-*

effect relationship stated in the strategic hypothesis. In chapter 1, I illustrated the idea of strategy as an *if-then* statement by examining a hypothetical everyday-low-pricing strategy by Wal-Mart, designed to drive an increase in market share: *If* prices are lowered storewide by an average of 2.5 percent, *then* market share in trading areas where we compete with Kmart will increase by at least 5 percent compared with stores without price reductions, over a period of twelve months. *Further,* gross margins can be retained at current average levels because logistics cost reductions equaling the lost gross margin can be achieved.

An important hidden assumption here is that Wal-Mart's retailing environment is no more than moderately variable. First, it is assumed that there will be only moderate variation in the results of this strategy across geographic locations, from Arlington, Virginia, to Stockton, California. And second, it is assumed that there will be only moderate variation in the results over time. Thus, for example, market share gained at the expense of Kmart over the first year the new strategy is in place won't be forfeited thereafter.

But life, as they say, is uncertain. Our discussion of the unraveling fabric of SCA in the last chapter suggests that, overall, strategic environments are becoming less stable and more variable. And our thinking about the degrees to which these environments are variable, using Sutherland's concepts and the preceding examples, reveals that there has been a pervasive loss of faith in the concept of *certainty* in business.

Now we can specify with some precision where seeking an SCA *is* appropriate: in deterministic environments, and in deterministic environments alone. For all other environments, we need something else—something that can buy back some of that lost sense of certainty. My suggestion is opportunity creation and exploration (OCE).

The OCE Cycle

The OCE concept is one of a cycle of activity rather than a single state of affairs to be achieved, as I've diagrammed in Figure 5-2.

1. Opportunity Creation and Discovery

The first phase of the OCE cycle is the rootstock, because the attitude of opportunity creation and discovery infuses the entire strategic effort. The idea is that in most environments—moderately to severely variable ones, as well as indeterminate ones—the appropriate approach is to generate numerous strategic hypotheses, test them, mod-

Figure 5-2. The OCE cycle has four distinct phases.

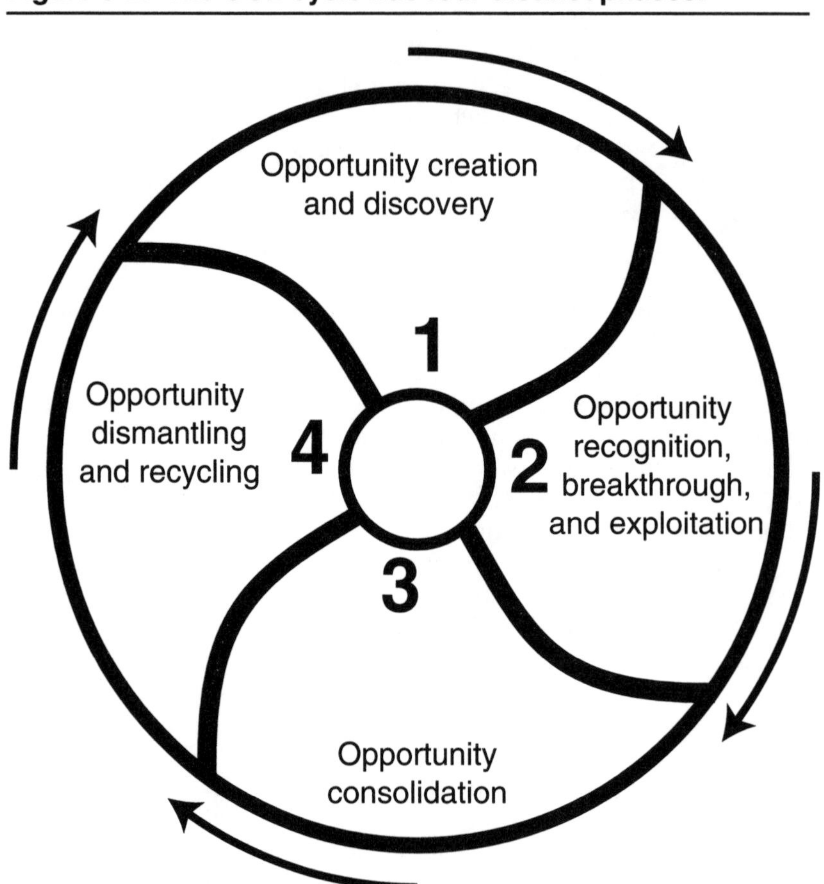

ify them to make them more precise, and winnow out the likely winners from the likely losers.

This attitude is the precise opposite of the typical strategic approach. Typically, a company and its management consultants seek the one perfect strategy for a company. At most, senior executives choose from a severely curtailed list of three or four options. Also typically, a company's planning staff or its consultants weight the evidence toward one or another of the alternatives to *confirm* the wisdom of that alternative. Usually, too, the planning staff or consultants look to an already existing fact base (for example, market research and cost data), rather than discovering new data and generating new tests that are appropriate for the evaluation of each and every *live* strategic alternative.

By contrast, what I'm suggesting is that a company's strategic planners (and its consultants) become a *laboratory of strategic experimentation and testing,* rather than an analysis factory bent on producing confirming evidence for a limited set of alternatives. This is necessary because the goal of SCA is unlikely to be achieved by following any single, long-term path. Instead, at any given time, a company has to create numerous options for itself. It must "let a thousand flowers bloom" to see which species and which varieties among them are hardy enough to withstand the gyrations of today's business markets.

This is particularly important in moderately and highly variable environments. By definition, they're environments where it's difficult to predict the results of strategic experiments by what *should be* the case. Only trial, test, and error can reveal causal links that can be depended on and how far they can be trusted.

A worthy model is that of a large pharmaceutical R&D company. Drug researchers learned long ago what strategists haven't. No matter how much confirming evidence there is that a new molecular compound *should* have this or that effect, nature offers no substitute for testing that compound, and often thousands of related ones, before yielding a small set of promising results. But nature usually permits extensive enough experimentation so that scientists can plot the probability curve relating cause to effect.

If you look up a pharmaceutical in *Physician's Desk Reference* or examine one of those drug interaction and side effects lists from a local pharmacy, you will find statements like "Dizziness, unusual fatigue, and sleeplessness were reported in 6 percent of those using this pharmaceutical." Good tests of strategic hypotheses can often provide a similar probability plot.

Nature offers no substitute for being precise. Nature permits no shortcuts. We shouldn't either, in pursuit of robust strategic hypotheses.

I'll have more to say about this first crucial stage in OCE a little later—particularly about where the hypotheses come from and how they're tested. But for now, let's sketch the other steps in the OCE cycle.

2. Opportunity Recognition, Breakthrough, and Exploitation

The second step in the cycle is obvious in concept but arduous in execution. One of the most difficult tasks of management is to recognize a success or something that can be turned into a success (often, "failed experiments" are really "near successes") among the numerous strategic hypotheses that have passed initial tests. This often means setting aside deeply rooted prejudices and putative *lessons from experience* and letting the facts speak for themselves, as much as they can.

Once this mental chasm has been crossed, the next item on management's agenda is the redirection of resources to create a breakthrough. A *breakthrough* is the production of the first measurable success from the strategic hypothesis. Organizationally, this translates into realigning people, money, and compensation to feed the new strategy—and starving less strategic initiatives.

For example, a project I recently participated in for a consumer electronics company showed that its most promising strategy required up to thirty new product introductions per year across half a dozen base technologies. This could be accomplished only if the company was willing to rechannel its engineers away from developing specific, singular devices for each market and variability, each with its own hardware. Instead, it needed a common hardware platform and the ability to toggle features on and off for different applications, using software. Management's job was to redirect R&D's people, projects, and attitudes toward this brand-new goal; gear up pilot manufacturing; and finally seek to generate sales in promising markets. Those first sales efforts, if successful, would constitute a breakthrough for a company that had been fixed on the idea of finding *the* perfect product.

From this, it's easy to see what I mean by *exploitation*: doing what it takes to enlarge on the just-recognized success of the strategy. In the example just cited, this would mean taking full advantage of the rapid-fire generation of new products by tweaking them to meet specific market needs, delivering them directly to new customers and markets, and using them to overshadow competitors' specialized offerings.

3. Opportunity Consolidation

The first two phases in the OCE cycle are primarily investment phases. There is little hope of gaining a return on investment (ROI) from opportunity creation and discovery; these are nearly pure investigation efforts. Only under very favorable and rare circumstances will a prototype product or service used to test a hypothesis in the marketplace return a profit from day one. Profits and a favorable ROI may be gleaned in the exploitation phase of stage two. However, a company is still intensifying its efforts to expand the successful strategy's application wherever it can, and this means higher year-over-year budgets. In the first two phases of OCE, the focus is on the strategy's offensive elements—on making the strategy's Hammer work.

The consolidation phase, in contrast, is where a proven strategy reaps its rewards. Investment drops, and resources allocated to the strategy's Hammer are stabilized or even reduced. For instance, in the consumer electronics company example, by the consolidation phase, relations with target customers—in this case, the main telephone service carriers—are well established, the company's R&D department is routinely churning out twenty to thirty new products per year, and revenues are fairly predictable.

Organizationally, the consolidation phase is marked by handing over the day-to-day strategic implementation work to eager but less experienced managers. Meanwhile, those who created and exploited the strategic opportunity prepare for the next phase.

Consolidation means more than resting on our strategic laurels, however. The intellectual challenge is to use the experience gained so far to refine the *if* and *then* portions of the strategy. The goal is to make links between *if* and *then* as definitive and as predictable as possible. For example, if we go back to Kmart's (hypothetical) strategy of upgraded stores in chapter 1, the consolidation phase would see a completion of the store-by-store upgrades and a refinement of upgrade techniques. (What is the best lighting level? Where and how large should the camera counter be? What signage works best?)

For strategic planners and thinkers, the consolidation phase is marked by the difference between *accuracy* and *precision.* Strategic hypothesis creation and testing focus on being accurate about what strategy will work best. In the consolidation phase, we have the luxury of becoming as precise as possible about the strategic details. This phase features particularly heavy data intensity, as spreadsheet jockeys analyze the results of the fully implemented strategy. These analysts have a field day in situations such as Kmart's, comparing one store with another, calculating same-store sales increases, figuring sales per square foot, and so on. But be aware that this kind of detailed analysis is useful only when a strategy has been thoroughly implemented. Otherwise, we run the risk of being extremely precise about the wrong strategy, using precision as a substitute for hypothesis generation and testing. That is, we can be precisely *wrong.*

The consolidation phase comes as close as the new millennium will allow to *normal business.* A mood of predictability and normalcy may pervade corporate headquarters. A new college graduate joining a company during this strategic phase might be forgiven for thinking that her projected career could be plotted with the precision of a satellite orbit.

Experience shows that this mood of complacent satisfaction can be as embracing and tenacious in a company as the tentacles of an octopus. The mood seems to be distributed by the very ventilation systems, even in once-successful companies that are losing millions today. Its emergence is a warning sign that management should move into the next phase of the strategic cycle.

4. Opportunity Dismantling and Recycling

The decline of SCA and the ascendance of inherent variability in business are signs that all good things must come to an end, even the most brilliant, well-tested, and thoroughly implemented strategies. The OCE concept suggests that instead of trying to build barriers around a successful strategy to keep it as viable as possible for as long as possible, we should proactively dismantle it. In this salvage operation, we take what we've learned and the assets we've accumulated in the service of the old and ruthlessly enlist them in the service of the new. Thus the stages of this phase are as follows:

- Continual testing of the current, successful strategy until its boundaries are revealed, anomalies begin to occur, and breakdown is seen just over the horizon.
- Identification (call it *accounting*) of all the assets—physical, financial, human, and emotional—committed to the strategy.
- Withdrawal of these assets from the old strategy—at a rate that can range from gradual to precipitous—and redirection of them to the support of the new strategy.

The testing of the current strategy—the attempt to falsify it—is never over. A company is continually seeking the boundaries of its truth. For example, in three years' time, the testing of the Kmart renovated-store program might reveal that (1) one of its boundaries is that it works only in communities where the average family income is between $28,000 and $60,000; and (2) even in stores within these communities, for health and beauty aids and casual wear, customers are increasingly turning to the membership-based warehouse clubs like Price, Costco, and Sam's. Over the horizon, management can see the end of its current strategy looming, as these warehouse clubs become the shopping venues of choice and Kmart's exclusivity wanes.

Some critics fault Borders Books and Music, the successful retail chain, for failing to see the boundaries and the end of its bricks-and-mortar strategy with the rise of Internet book sales.[6] Borders built its success on an inviting store format, including coffee shops and special events; a large and current book selection; and aggressive negotiation with distributors and publishers for return rights. The analog to the hypothetical Kmart strategy is almost exact. Research could have shown that just over the horizon lurked Internet book shopping.

The next step, accounting for the assets involved in the current strategy, is required so that those assets can be deployed as Pivots for a new strategy, directed in the service of a replacement strategy, *or* mercilessly liquidated. Seeing the looming end for its current strategy,

Kmart might shutter, relocate, or withdraw management talent from those stores operating outside the newly discovered boundaries of the strategy and those in trading areas likely to be dominated by the warehouse clubs. Those freed resources might then be better deployed in a new strategy—Kmart's own lower-priced membership club. The same observations apply, mutatis mutandis, to Borders.

The Cycle and the Spiral

Perhaps you've noticed that the OCE cycle has become a spiral. The internal logic of strategy as falsifiable hypothesis and the realization that most business environments are inherently variable over time lead us, as servants of logic and masters of prudence, to

- continual testing of the current strategic hypothesis and
- continual opportunity creation and discovery.

Because it's often too late to start opportunity exploration when your current strategy is winding down, it's better to start the quest and keep it going during the consolidation phase of a successful strategy and, if you are bold enough (and have the intellectual resources), during the breakthrough and exploitation phases. Successful drug companies that have found a winning compound don't shut off the gas to new research efforts, even in the same therapeutic area (say, diabetes or depression). The world is moving too fast for even the most successful companies to stop pushing their strategic function. Thus, the cycle becomes a spiral, or actually a *cycloid,* to mathematicians (see Figure 5-3).

Application of OCE to the Diversity of Business Environments

It would be the pinnacle of irony if OCE, with its emphasis on diversity and variance, were meant to be applied in cookie-cutter fashion to every business environment. Fortunately, such is not the case. Rather, OCE is meant to be applied in a manner consistent with the diversity of business situations. Although no book can summarize every business environment, we can use Sutherland's buckets of degrees of variability to segment the business world and thus the ways in which OCE is most appropriately applied (see Figure 5-4):

In Figure 5-4, I have assigned coordinates because some of the

Figure 5-3. Recycling older strategies into new creates an OCE cycloid.

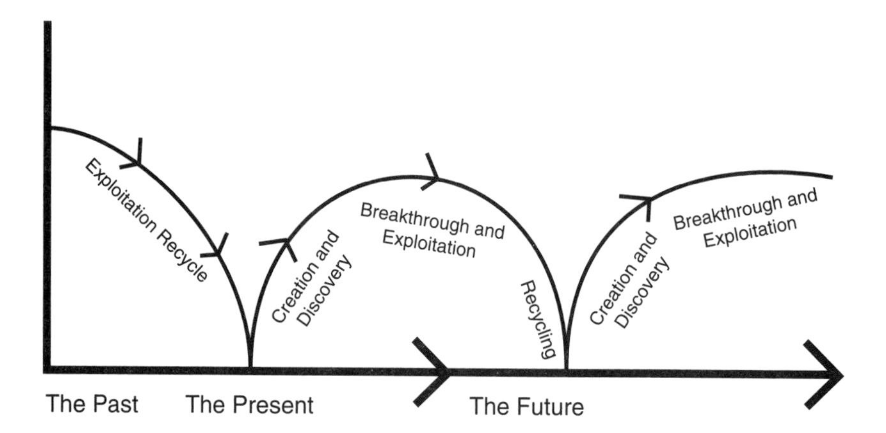

Figure 5-4. Sutherland's business environment buckets illuminate characteristics of strategic hypotheses during phases of the OCE cycle.

	Column A Determinate	*Column B Moderately Variable*	*Column C Severely Variable*	*Column D Indeterminate*
1. Opportunity Creation and Discovery	One right answer; SCA possible	Strategic hypotheses are variations on a few themes	Need highly diverse set of hypotheses	Very high number of hypotheses; testing and launch virtually indistinguishable
2. Breakthrough and Exploitation	Invest heavily in one right answer	Invest moderately in several avenues	Reinforce apparent success	Shut down losing approaches quickly
3. Consolidation	Erect barriers; exploit monopoly-like conditions	Reinforce success without long-term resource commitment	Use testing to try to move to moderately variable; keep assets mobile	This phase is truncated—harvest is reaped during exploitation phase
4. Dismantling and Recycling	Slow, methodical withdrawal of resources to a related or similar business	Begin dismantling when testing shows probable falloff in ROI	Organize for rapid redeployment of assets at all times	Dismantling begins with peak of exploitation phase

cells are worth commenting on. First, if we compare columns A and D, especially in rows 3 and 4, we see the vast gulf between the business environment of cells 3A and 4A, which lend themselves to the erection of a more or less sustainable competitive advantage, and those of 3D and 4D, where even the OCE cycle is compressed.

Furthermore, an analysis of changing business environments leads us to believe that most of them are moderately to severely variable (columns B and C). Within these columns, however, experience suggests that most companies have the following three very understandable tendencies:

1. *They lose their willingness to entertain new hypotheses (cells 3B and 3C).* So many entrepreneurs have the attitude, I'll try anything once, to see whether it works. This is partly the result of having had enough success to move into the exploitation, breakthrough, and consolidation phases. It's natural to think, That was then; this is now. Let's get on with making some money.

I see this syndrome at work with successful executives who are recruited to work their magic in a new firm. Instead of going into Company B with the wide-open attitude that made them successful at Company A, they often figure that they have mastered the art of success and merely need to emulate what they did before in the new situation. This is reminiscent of Captain Queeg's insistence during the ice-cream theft incident in *The Caine Mutiny* that a wax impression of a stolen key must have been made, since that's what he had uncovered in a similar situation many years before.

2. *They cling to outdated strategies and their underlying hypotheses (row 3, especially cells 3B and 3C).* An extended but instructive example of this is occurring now with health maintenance organizations (HMOs). The original strategic concept, especially among staff-plan HMOs (in which physicians are employees of or are closely tied to the health plan), was really a cost-based, economies-of-scale approach that Henry Ford would have instantaneously acclaimed. The idea was to create savings by consolidating overhead functions formerly distributed among many physicians' offices, gain lower sourcing costs—especially for pharmaceuticals and specialists' services—through volume purchasing power, and scrutinize participating physicians' practices to reduce wasted time and unnecessary procedures and prescriptions. The public relations spiel and the government relations angle was that HMOs would work to maintain health by preventing disease in the first place rather than treating it afterward.

Although group health organizations of one type or another had been around since the 1940s, they became enshrined in federal law in the 1970s. By the 1980s, their time had come. Rising overall medical costs had become an annoyingly large percentage of total costs for some large companies. In 1996, for example, Ford and General Motors estimated their health care costs per car at $510 and $1,200 respectively.[7]

By the early 1990s, it was becoming clear that the cornerstone hypothesis of the HMO edifice—that the practice of medicine had scale economies—had run its course. HMOs were not unsuccessful, but by 1993 or so, all the low-hanging cost-savings fruit had been plucked, and plan buyers (large corporations, especially) had been very successful in negotiating favorable pricing on premiums. The balance of power had shifted to the large employers because they could deliver the high volume of members (their employees) the HMOs needed to cover high fixed costs.

Yet the larger health plan companies are still seeking those elusive scale economies. If they can't create them, they'll buy them. For example, in the late 1990s, giant Aetna merged with U.S. Healthcare and bought Prudential's health insurance business. The Blue Cross and Blue Shield plans are forming holding companies such as Premera, Anthem, Horizon Health, and Amerihealth, so that they can circumvent the Blue Cross association–franchise rules that limit their licenses to doing business within a state. Now they can lose money across state lines.

Meanwhile, the HMOs' overlooked siblings, preferred provider organizations (PPOs), are increasing in popularity and making money.[8] PPOs are simpler contracting arrangements in which health insurance companies negotiate with selected providers for lower medical costs, in return for steering volume their way. PPOs are more flexible, and there is no pretense about overhead cost reductions or about "maintaining health by preventing disease." There's a successful strategy lurking in there somewhere.

All in all, then, companies tend to stick with a strategic hypothesis too long, driving by looking in the rearview mirror. Too frequently, it takes a crisis to provoke new thinking. And once the crisis begins, management often wants a miraculous strategic plan (and usually the *one* right answer) that will deliver consolidation-phase profitability—all because there isn't world enough or time enough to go through the creation and discovery effort and the exploitation and breakthrough effort that led to success in the first place.

3. *They permit (or force) strategic Hammers to evolve into Pivots.* During the consolidation phase, management stabilizes and seeks to make efficient and profitable its current successful strategy. It even begins withdrawing some resources. The current strategic focus goes on de-

fense while resources are realigned for the new strategy's Hammer. This process accelerates in the dismantling and recycling phase, where the current strategy becomes the Pivot for support of newly developed strategic aims.

Letting a Thousand Flowers Bloom

Stress has become the byword of people everywhere in the industrialized world. And today, the word is apt for most companies in the environment of global competition and rapid change. They are stressed.

Recent research shows that the responses of plants and animals to stressful environments are very much like OCE's first phase of opportunity creation and discovery and occur far more rapidly than previously thought. As evolution progresses, species become more and more adapted—specialized—to exploit their environments. As with companies, the more perfectly adapted a species is to an ecological niche, the more defensible it is from would-be encroachers. Observers of history have not overlooked the analogy to corporations finding niches in which they can be successful.

In classic evolutionary theory, species change to adapt themselves to ecological niches through random variation and sexual selection. But this adaptation, which eventually gives rise to divisions in species and the genesis of new species, is slow and incremental. Charles Darwin and his followers believed it to be far too slow to observe in the life span of a human scientist.

But recent work by Darwin's descendants has shaken this view to its roots. Field observation has shown that change can, in fact, occur rapidly within a species. New species, not just variations within species, can arise within a handful of generations. And two broad phenomena foment and enable this rapid evolution: environmental stress and hybridization. Both have strong analogs in today's economy. Both flash warning signs for strategies built on SCA.

In his exciting book *The Beak of the Finch,* Jonathan Weiner describes how researchers Peter and Rosemary Grant (and others) traced the evolution—or, rather, the revolution—of variations among finches when their habitat, on isolated islands in the Galápagos archipelago, suffers from such periodic calamities as droughts and storms. The finches are caught between times of poverty and plenty. In times of drought, diminished food supplies and increased competition from other species result in widespread starvation. Sexual selection takes its toll as well. Some survivors simply aren't attractive enough—or quick enough—to find mates.[9]

The first, most direct force for the development of new species is environmental stress, particularly as produced by major climatic events such as El Niño or drought. The Grants showed that environmental stress is enough to cause species like the finches to subdivide within a period of years, not decades, centuries, or millennia. Finch variants become more and more adapted, for example, to utilizing specific foods and sources of food and to taking advantage of new types of cover and protection from predators. Such adaptations are often rooted in climatic or weather changes. Research now shows that these variants can quickly become more differentiated, and when variants are so differentiated that they do not or cannot interbreed, they deserve to be called different species.

The same process of rapid specialization and increased variation takes place among bacteria and in maize, as these organisms respond to stressful environments.[10] In sum, nature's response to stress is the creation of many variations on a single theme. These variants are clearly *living hypotheses* that some adaptation or another will boost the individuals' chances for survival.

My suggestion is that in these times of global competition and rapid change in the business climate, industries and companies should emulate nature by creating and trying out many strategic hypotheses. Many of these hypotheses will be thoroughly (and deservedly) falsified, but the survivors will hold the key to success.

The other important force that instigates change in species and has a strong analog in the business world is hybridization. In biology, hybridization is the mixing of genes from related but different species. It was once thought to be a source of variation in the plant kingdom exclusively, but it is now known to be a general engine for the origination of species. "Hybridization provides favorable conditions for major and rapid evolution to occur," wrote the Grants in the journal *Science.*[11] And such hybridization, though confined to a small minority of bird species (6 to 10 percent), may be necessary for survival: "So the birds that are bound to this little island, breeding where their line has bred for generations, may often need as great an infusion of fresh variation as plants whose seed drift hundreds of miles on the wind."[12]

I would argue that the advent of fast, cheap transportation and fast, cheap communication has boosted not only the competitive environment but also the chances for and the fact of business hybridization. American companies import the idea of *kaizen,* or continuous improvement, from Japan, while they export the idea of *pay for performance* to France and Germany.

Cross-border mergers can become true hybrids. The theory of the Daimler-Chrysler merger was, for example, to blend Chrysler's peculiarly American styling and marketing expertise with Daimler's quality control, engineering, and purchasing savvy. Most hybrids in nature are unsuccessful or sterile. It remains to be seen whether Daimler-Chrysler's hybridization will succeed. The likelihood is that with the Germans asserting more and more control, it won't be able to hold on to the American "genes" that might make the hybrid successful.

Putting Sutherland and OCE to Work

Putting OCE to work is not a task for a day or a week. It is a process of creating a realization that a corporate strategy is a living thing that has a life cycle. As a living thing, it must accommodate, shape, and respond to its environment.

The first thing to do is to make an inventory of your current and future business environments—the broad ecology of your business situation—according to Sutherland's buckets. Here's one way to start thinking about it:

1. If I make a change in prices for my product, how well can I predict the effect on volume, customers' reactions, and competitors' reactions?
   a. Very well.
   b. Pretty well, within limits.
   c. Broadly speaking, but without precision.
   d. I can only make an educated guess, but you never know.

2. Considering how much warning time I usually have, how well can I predict new product introductions from competitors?
   a. Very well.
   b. Pretty well, within limits.
   c. Broadly speaking, but without precision.
   d. I can only make an educated guess, but you never know.

Now perform the same analysis for all the other main components of your business environment—for example, the stability of your supplier and customer populations and relationships, crucial talent, intellectual property, and key assets. Review your answers. More likely than not, a pattern will emerge that indicates which of Sutherland's buckets your business is operating in.

3. The most important question to ask yourself is, Is our corporate strategy getting stale? Symptoms of staleness are:

- Complacency of senior and middle management about their jobs

- Stagnant or declining ROI
- Stagnant or declining returns on managerial effort
- Increasing competitiveness in bidding for customers or market share
- A history of disappointment in recent strategic moves
- A deep belief that recent strategic moves have been futile

A positive answer suggests that you should immediately begin the opportunity creation and discovery phase of OCE. As that new strategy comes into focus, it won't be too late to start concurrently the last phase, strategy dismantling and recycling: the conversion of key elements of the current strategy to the defensive. Identifying the defensive Pivot and the Bearing in the Pivot is of major importance, while your eyes are understandably scanning for a glimpse of the Hammer and the Hammerhead in the new strategy.

4. If you conclude that your strategy isn't stale, the next task is to identify what phase of the life cycle your business strategy is in. What cell of Figure 5-4 is your company currently occupying? The answer will help you look ahead to the next stage of the OCE cycle. Once you establish where you are in the strategic life cycle, seize control of your strategy. If you don't seize control of it, your strategy's life cycle will accelerate of its own accord, and you'll find yourself in strategic crisis.

The remainder of this book is about seizing control of strategy—implementing strategy thoroughly and aggressively.

Notes

1. *Business Week,* 12 January 1998, 41.
2. Diane Brady, "Xerox," *Business Week,* 12 April 1999, 92–100.
3. Peter Elkind, "Viva Zapata? Mmmm . . . No," *Fortune,* 17 August 1998, 25–26.
4. John W. Sutherland, *Systems: Analysis, Administration and Architecture* (New York: Van Nostrand, 1981).
5. Sutherland uses the more technical term *stochastic* to describe these categories.
6. Anya Sacharow, "Book Selling's Online Jungle: Borders Dares Amazon, B&N," *Brandweek,* 27 April 1998, 34.
7. "GM Treats Health Providers Like Any Other Suppliers," *Wall Street Journal,* 9 December 1996.
8. Allison Bell, "PPOs Still Lead HMOs in U.S. Market," *National Underwriter,* 19 October 1998, 32.

9. Jonathan Weiner, *The Beak of the Finch: A Story of Evolution in Our Time* (New York: Vintage Press, 1995).
10. Ibid., 221–23.
11. As quoted in Weiner, 198.
12. Ibid., 198, 200.

Chapter
Six

Better Tools for Better Strategies: Creating, Testing, and Falsifying Strategic Ideas

Nothing is more dangerous than an idea when it is the only one you have.

—Emile Chartier[1]

Once, long ago, I made the mistake of responding to a philosophy professor's uncomfortably penetrating question with, "I want to keep an open mind." His reply was, "The only point of having an open mind is that at some point you can close it." His point was well taken: It is indeed a person of luxury and wealth who can afford to keep an open mind forever. In business, executives have that opportunity but seldom.

This book is about how businesspeople can keep open minds, and when and how they should close them. In that sense, it is a book about moving mountains—that is, an orderly way of changing minds. The first stage of the opportunity creation and exploitation (OCE) cycle consists essentially of opening your mind (generating strategic hypotheses) and then closing it judiciously (through the testing and falsification of those hypotheses).

This chapter provides tools and attitudes for opening your company to strategic hypotheses or opportunities and then choosing among them. I want to emphasize attitude, because there is no royal road, no cookbook, to strategic success. The only ingredient that is unquestionably necessary is an inquiring mind. It's helpful, of course, for that inquiring mind to have good tools. There are dozens of these, and although this book is no place for an encyclopedic compilation of them, it is useful here to highlight some of the most valuable tools and critique their usage.

Attitudes and Tools for Hypothesis Generation

In a world of fierce competition, the most important criterion for hypothesis generation is beating your competition to the best ideas. To do this, you must

- Have lots of strategic hypotheses per unit of time—be intellectually productive.
- Have a broad range of strategic hypotheses.
- Get your strategic hypotheses (your good ideas) from the right sources.

This may seem simple, but experience suggests that in many companies, *just the opposite* is the case. Most companies and other organizations handle ideas in a way that *suppresses* idea productivity. For instance, ideas are often created and written up, then judged, accepted, rejected, modified, and funded, one at a time or in small batches. As a result, they get clogged in the pipeline. This means that *only as many ideas as will fit in the judgment and approval bottleneck* are available to a company.

Another problem is that the same people generate ideas again and again. They are usually top management, R&D or product development, or sales people in contact with customers. Sometimes they're marketing people; now and then, they're outsiders such as advertising agencies or consultants. This range of sources is not very broad, it may not produce a wide range of ideas, and it may not even include the right sources. People in R&D and sales often don't realize that their new product ideas or selling ideas might require a wholesale change in corporate strategy. That's one reason they get shot down so often.

Nearly every book on creativity will tell you that the way to find a *good* idea is to start with *a lot* of ideas. Creativity guru Roger von Oech provides an apt example: "Edison knew 1800 ways not to build a light bulb."[2]

I believe that the same principle applies to creating strategic hypotheses. After all, these are nothing but tentative assertions about what will happen if certain conditions are fulfilled. Inventing strategic ideas has a lot in common with other forms of invention—for example, coming up with new uses for toothbrushes, new ways to make toothbrushes, and new ways to sell toothbrushes.

But strategic thinking faces special barriers. For one thing, in strategic thinking, the stakes are higher for your company and, not incidentally, for your career. For another, strategy is abstract: You can twiddle, bend, break, or boil a toothbrush or even an ad for a toothbrush, but it takes an act of imagination to bend—not to mention torture—a strategic hypothesis.

For McDonald's: If McDonald's is the place for kids, what is the place for seniors? For house pets? For Microsoft: What if we split into two companies—the Windows operating system company and the applications company—or even three? It didn't hurt Lucent much to be split from AT&T—far from it. For Motorola: What if we acquired a telephone network company? A fiber optics company? Wireless phones are dropping yearly in price. Why fight it? We could give them away, as long as they're connected to our network.

It's important to overcome strategic thinking barriers, because the competition might get to the next good idea first. As of this writing, free personal computers are just beginning to enter the market, linked to mandatory viewing of advertising, exploitation of the user's demographics, or multiyear Internet subscriptions. Bill Gross's IdeaLabs was the first to launch the idea to a broad public. But three months earlier, my company suggested the idea to a large, successful personal computer manufacturing company. It was highly intrigued and invited my team to Texas, but the executives couldn't find time on their calendars for us all to get together. Now a Korean-funded upstart, eMachines, is in the game, linked with the CompuServe unit of America Online (AOL). The folks at AOL are awake and don't care about the cost of the machines their subscribers use—as long as they use AOL. Tiger Direct, a low-ball clone maker, is also giving this idea a shot.

I'll stick my neck out: By the time this book makes it to print, free PCs with subscription links and arrangements with preferred retailers will be at least as common as free cell phones and almost as common as teaser-rate credit card solicitations. Where will my Texas friends be then?

If the initial goal for hypothesis generation and opportunity creation is having lots of ideas faster than the competition, then having some help finding those ideas would be nice. This is easy: There are people who specialize in providing that sort of help. Plus, you get an expense-paid trip to Menlo Park or Hawaii or Rio or Salzburg—where, it seems, the good ideas (and the gurus) are found.

But I'm going to warn you about the tools that are as close as the address book you're likely to reach for first. Just as you are sometimes tempted to use a nearby paperweight to fix a computer but (usually) don't do it, most companies are tempted to use two tools to help them generate strategic ideas that they probably shouldn't: market research companies and consultants.

The Trouble with Market Research

In my work helping companies develop strategies, I'm almost invariably presented on day one with a bulky market research report. Usually it's a huge compilation of the results of a massive phone or mail survey, and often it's very well done. But the senior executives are dissatisfied.

The reason for their dissatisfaction is one of the things that make market research so appealing in the first place: It is quantitative. As a person whose motto is "The numbers talk to me," I have nothing against quantitative research, but not at this stage.

To get the numbers, market researchers have to ask questions that are:

- Limited by the vehicle of response—written questionnaire, phone call, or Internet survey.
- Limited in the potential range of responses—the infamous forced categorization of the phrase "Which one of the following best describes . . ." in order to obtain adequate numbers of responses per category.
- Limited in number—limited by the patience of the respondent and by the (often questionable) intelligence of the interviewer.
- Usually based on respondents' past experience—because a format-bound phone call or written survey is no place for exploring hitherto unknown needs or desires.

From this mass of numbers, even if they're summarized in cute bar graphs and pie charts, it's often difficult to extract any hypothesis. This would come as no surprise to practicing scientists, who know that it is better to come to data with an idea or a hunch first and then send in the bloodhounds.

Further, to discover what is really on people's minds, market researchers would have to be telepathic or clairvoyant, rather than slaves to a questionnaire script. That's because, as researcher James Billington puts it, getting ideas from the market is a matter of active, absorbed observation, not digital questioning. He illustrates the difference between observation and inquiry this way:

> The customer communicates with you all the time, whether or not you are aware of it. Paying particular attention to unspoken cues can mean the difference between success and failure.
>
> In "Spark Innovation Through Empathic Design," Dorothy Leonard and Jeffrey Rayport argue that customers

> often cannot articulate their needs, even though they unconsciously provide examples of those needs all the time. Observation thus becomes more important than inquiry, the traditional means of market research. . . .
>
> When Nissan Design's president, Jerry Hirshberg, noticed a couple on the side of the road trying to jam a couch into their minivan, he stopped and spoke with them. They said they had bought the minivan because of its spaciousness, but discovered they had to take out the seats to fit anything large into it. Observation led to inquiry: this conversation sparked Hirshberg's idea of using six-foot runners so that the van's back seats could fold up to make room for cargo.[3]

The discovery of strategic hypotheses isn't identical with the development of new product ideas, but new products and new ways of marketing and distributing them are often at the heart of blockbuster strategies. It pays to pay attention to the wisdom of product developers.

Another lesson from scientists is the role of serendipity, chance, and just plain good luck in finding good ideas. There is a long list of discoveries and inventions that saw the light of day when scientists and inventors looking for one particular thing discovered another. For example, the world's chemical industry came into being when William Henry Perkin was looking for a way to synthesize quinine but instead discovered aniline blue, the first artificial dye made from coal.[4] A famous case mentioned earlier is Alexander Fleming's discovery of penicillin while in search of a cure for influenza.

According to business researcher James Brian Quinn:

> Leo Baeklund was looking for a synthetic shellac when he found Bakelite and started the modern plastic industry. At Syntex, researchers were not looking for an oral contraceptive when they created 19-norprogesterone, the precursor to the active ingredient in half of all contraceptive pills. And the microcomputer was born because Intel's Ted Hoff "happened" to work on a complex calculator just when Digital Equipment Corporation's PDP8 architecture was fresh in his mind.[5]

The straitjacket of quantifying market research immediately rules out the opportunity to experience the benign effects of chance on the creative strategic process. It has its place—and an important one, as we'll see later in this chapter—but it isn't *here,* at the idea creation phase.

The Trouble with Consultants

If many companies rely too much on market research, it is also true that many companies place too much trust in consultants. I have no objection in principle to using consulting firms; indeed, it's often a good idea. The best consultants are usually smart, competitive, industrious, eager to please, dispassionate, able to access their own sources to obtain information it might take you months to acquire, and ready to serve as a resource dedicated to a problem—when your smartest, most energetic staff is busy making *today* happen.

Most companies select consultants for major projects in a competitive *shoot-out* or *beauty contest.* This too is fine, as far as it goes, although in my experience, the major firms are all competent at uncovering the most important issues, running the financial numbers, and coming up with defensible solutions.

It's the way consultants are used that needs thorough reform in many companies. If you've been following my argument about the need for the falsification of strategic hypotheses, you'll recognize that obtaining a consultant's services is just the beginning. Being human, consultants, like the rest of us, will bring forward only what supports their brilliant strategic ideas. They'll overlook, dismiss, or belittle the evidence that doesn't support their ideas. They want to get paid for having brilliant insight, and after weeks and months of late nights, they have an emotional commitment to their answers. Would you want it any other way?

Thus, at the end of the typical consulting engagement, the client has a very strong hypothesis or two about the strategy it should undertake. But most often, both client and consultant mistake this point for the end of the strategic work. Everyone is anxious to get on with *implementation.* But my argument should make it clear to you that this is where the heavy lifting begins. I pick up the theme of how to use consultants in the testing and falsification stage later in this chapter.

Better Tools for Hypothesis Generation

The Executive in the Parking Lot

The most important tool for hypothesis generation is the energy and time of senior executives. Not only are these executives usually smart, aggressive, experienced, and knowledgeable, but they own the strategic problem more than anyone else does.

The problem, of course, is to get them out of their offices and into the front lines, where they can see and feel how today's strategy is faring in the marketplace. But it's worth the effort. In my experience, the floodgates of executive creativity open up when we, as consultants, bring them face-to-face with customers. When I can't do this, I find it very effective to bring tapes or even quotations in cold print to their attention. There seems to be nothing more stimulating or credible than real—or even *virtual*—visits to the field.

Research bears this out. Over 70 percent of successful new product ideas originate with market needs, and 50 percent involve direct input from customers, according to Billington.[6] Further, research (as well as common sense) suggests that highly motivated, interested (not to mention *self-interested*) executives can uncover market needs that customers can't articulate. The example I cited about Nissan's president stopping by the side of the road to talk with frustrated minivan buyers illustrates the impact executive eyes can have on a company's decisions.

At GM, executives are required to buy or lease a GM product bought and serviced in special GM offices and garages attached to their office buildings. By contrast at one Japanese company's southern California headquarters, marketing executives buy several competitors' cars each year from real dealers at company cost, all for their personal use. Who is likely to gain better marketplace intelligence?[7]

The other advantage to getting executives out of their offices and into the malls, parking lots, and economy-class cabins of airliners is that when they get home, their experiences turn into ideas. And those ideas turn into hypotheses much faster than ideas that filter up, memo by memo, from field sales engineers and night watchmen.

The Focus Group

Focus groups have been around for decades. Experience indicates, however, that we expect both too much and too little from them.

We expect them to be good testing grounds for strategic hypotheses or new product ideas. But if a focus group or two pan an idea, it's best to look again. And if they laud it, that's not confirmation; in fact, it's almost no information at all. No group of fifteen to twenty people is a statistically valid sample. Nor are five such groups.

Additionally, it's important to remember that focus group respondents, for all their good intentions, have *no stake* in what's being presented to them. Whereas executives have almost too much of a stake—a job, reputation, a bonus—focus group respondents aren't even typically paying customers for the proposed product, the market entry strategy, or the value proposition we're testing. At most, they're *prospective* customers.

Finally, although focus groups are great at reacting to products and service ideas, they are usually ordinary folks or potential purchasers. They don't think strategically about competition, about timing, about resources, or about technical feasibility.

The bottom line is that you should use focus groups to help:

- Generate ideas.
- Push your thinking.
- Rule out *nonstarter* hypotheses.
- Provide raw material for generating strategic ideas.

Don't use them to:

- Test or falsify *live* hypotheses.
- Generate fully cooked strategies.
- Validate features, benefits, advertising messages, or the like.

We also expect too little from focus groups, because most of them produce only a third to a fifth of the insight they are capable of. An ordinary focus group, even with the most skilled moderator, is usually dominated by two or three people. Further, only one person can speak at a time (and be understood). Finally, most focus group meeting environments are not conducive to bringing out the best the participants have to offer about their private responses to ideas. We are, after all, drawing on their past experience and asking them to admit us to the real world in which they live. I have had success using three tools to overcome these defects.

First, for provoking and recording reactions to many new products or market entry strategy situations, I prefer to set up an electronic boardroom and use group decision software. An electronic boardroom contains fifteen or twenty computer workstations with keyboards and monitors. The computers are networked to group decision software. Two or three oversized screens hooked up to the network and the Internet complete the setup.

It's the group decision software that provides the key to this group's productivity. When the moderator asks a question—often using visuals on the oversized screens—the respondents type their comments at their workstations. This allows everyone to "talk" *at once:* no more waiting to take turns to speak, and no more domination by the two or three loudmouths you find in every group.

Equally important, comments are anonymous (although they

need not be). That strips away the self-censoring that hierarchy promotes. Anonymous comments can be displayed on the overhead screen for everyone to react to. That provides an atmosphere for better questions and better answers.

In one session, a group of U.S. Navy and Marine Corps officers ranging in rank from rear admiral to captain were thrashing out the strategy for an opposed beach landing. An anonymous comment posted on the public screen enraged one of the generals with its impertinence. "Who made that [expletive] comment?" he stormed. There was dead silence. The civilian moderator took the general by the arm into the hallway, with the admonition, "General! That's why you contracted for an electronic boardroom session. Isn't this just what you wanted?" The general bit his lip and calmed down. "You're right. Even the idiots have a point now and then." The impertinent comment turned out to be a turning point for the whole discussion of landing strategy. In the end, the session created a far better strategy than the textbook war college plans the participants had all brought to the session—one more economical of both time and lives.

The electronic boardroom atmosphere also cuts through less formal social hierarchies. I recently led a focus group concentrating on the best way to introduce a new diagnostic test into doctors' offices. The brand, quantity, and format of this particular test for a fairly routine condition are generally specified by the laboratory manager or head nurse of the physician's practice. This focus group included doctors, nurses, and lab technicians. We confirmed electronically what we had suspected: that the physician is typically ignorant of the test type, format, and price. He or she is usually aware only of brand, at the very most.

But when we turned off the software for a regular focus group brainstorming session, the *physician–nurse–lab tech* pecking order reasserted itself, even though these people had never met before and probably would never meet again. Two doctors dominated the whole discussion. One repeatedly demanded that "the test should be as easy as testing the pH in my swimming pool." Another, just out of residency, meekly argued with him. Meanwhile, the true purchasing decision makers sat in stony silence. If we had simply used this traditional output, we would have had a very skewed idea of the best strategy for introducing this new product.

It's no exaggeration to say that electronic boardrooms permit three to five times the productivity of typical focus groups. Since everyone can comment at once, and since every respondent can choose to be anonymous, it's not unusual to have a *transcript* of over a hundred pages of respondent commentary from a session, compared with twenty to thirty pages from an ordinary focus group. Figure 6-1 shows

Figure 6-1. Electronic boardroom solves many common focus group problems.

Problems	*Electronic Boardroom Solutions*
▪ Time restrictions: 120–150 minutes restrict subject matter ▪ Only one person can talk at a time ▪ Verbal views often modified for politeness ▪ Views of quiet, less vocal, or intimidated participants not heard ▪ Output is always verbal, not quantitative ▪ Output is tedious to review -Video and audio tapes -Filtered, biased interpretations -Written transcripts that are time consuming to read ▪ Observers often hear "confirmation" of preexisting hypotheses	▪ Simultaneous input from all participants ▪ Text input requires organization of thoughts ▪ Anonymous text input allows for frank—often brutally frank—views ▪ "Group-think" is avoided ▪ Output is available in word processing and spreadsheet format almost immediately ▪ Output can be verbal or quantitative ▪ Output can be easily organized by topic

how the electronic boardroom method can solve focus group problems.

Second, focus groups can be made more productive as idea generators if you provide respondents with lots of props. They don't have to be fancy, but they have to be visual. For example, I recently helped a number of companies launch Internet strategies. The most productive approach has been to provide numerous simple, highly graphic mock-up HTML-like pages suggesting different features for the companies' new Web sites. Respondents quickly understand that the concept has been merely sketched out.

A particularly productive subspecies of this approach is simulated shopping experiences. Here, the props are products that prospective purchasers might like to buy. For example, a cell phone company I've worked with arranged a spectacularly successful set of focus groups worldwide by laying out mock-ups complete with descriptions, price points, and alternative brand names. It then provided respondents with simulated cash to purchase the cell phones. From respondents' reactions—particularly how they allocated their money across cell phone features and brands—the company was able to generate a rich set of hypotheses to flesh out and test.

Third, a technique called *guided imagery* can be a powerful way to overcome the inertia every moderator feels when he or she is hoping to ignite a creative, blockbusting brainstorming session. That word *inertia* is no mere metaphor: When minds think for months and days on end about problems, they naturally develop habits of thought about those problems. These habits get reinforced and become incrementally stronger every time they are tapped.

Guided imagery is a technique for getting respondents to jump the railroad tracks of thought. It's derived from work by George M. Prince and W. J. J. Gordon.[8] Their research indicates that it's often productive to take focus group respondents on a mental detour when seeking new ideas. My experience suggests that this is right on target.

Guided imagery is especially useful when you are trying to generate hypotheses in such tight corners as the following:

- You need to know what your colleagues think would be the ideal solution to a strategic problem, even though everyone feels the pressure of current time, money, and competitive constraints.
- You need to shape a value proposition for the introduction of a new product, retail concept, or service.
- Your market share is stagnant, and you need to find new customers for your product or service.

Guided imagery is one of the most intriguing, unbusinesslike business techniques you will ever use. It requires the focus group moderator to lead the group on a fantasy excursion—one that, at first blush (or even second), has absolutely nothing to do with the problem at hand. As a kind of mental play, this technique appears to free up the mind from normal paths of thought.

One of my firm's clients was recently looking for something to make its new Internet site truly exceptional in an increasingly crowded field. My firm led the group on a guided-imagery excursion using the history of the compass, from the initial discovery of the lodestone in ancient China, through Marco Polo's introduction of the compass to the West as a curiosity, to its sudden importance to navigation during the Great Age of Exploration in the fifteenth and sixteenth centuries. We then asked the group to describe the features of the ideal compass if they were modern-day explorers. You could see relaxed attention permeate the group as a playful mood took over the room. The replies came slowly, then faster and faster. We then asked how they would compare the ideal compass to the ideal Web site for the client. The result was what the client called a "killer application" resulting in a whole new product offering for its customers—and a profitable revenue stream.

Here's a typical electronic boardroom focus group program when the objective is to generate new ideas for strategic hypotheses:

↓ Introduction
↓ Overview of situation, product, or service: quantitative and qualitative commentary
↓ Shopping excursion with competitive comparisons: quantitative and qualitative commentary
↓ Preparation for brainstorming session: guided-imagery excursion
↓ Brainstorming session

Although it's theatrically risky and you can feel foolish leading a talk about compasses or princesses and dragons or undersea travel, there is nothing more rewarding than having focus group respondents insist on talking about their new insights long past their (and your) bedtime.

The Stretch Target

This tool is an oldie but a goodie: Create an apparently impossible target in market share, profitability, product features, productivity, cost, or whatever. Then challenge your team or your consultants to imagine any way possible to meet the goal.

Humility is not a virtue normally associated with successful businesspeople, but occasional humbling experiences can be a valuable inoculation against pride. So it was recently, for me and for the consulting team I was assisting. Our task was to formulate a five-year strategy for a consumer electronics division of a large Asian conglomerate. About halfway through the project, our team perceived that our realistic suggestions about how this number-seven player might become a number-five player in its market were not meeting the client's expectations. At a two-day meeting in a dimly lit, overcrowded hotel conference room, it finally came out that the company wanted a strategy that would make it number one (or, at worst, number two) in the market in five years.

"Silly" would be the politest adjective to describe this goal. Nevertheless, days of argument and wrangling wouldn't shake the client. It was a top management imperative. Over the next few days, to our own astonishment, our team rose to the challenge of figuring out just what it would take to accomplish this impossible dream.

These insights did not change our hearts or minds about getting the client to the number-one spot. But they produced several important outcomes: They brought out in sharp relief the competitive and resource barriers to the goal. They produced four new ways to gain market share that we hadn't previously thought of. They made us think through the steps required to make *any* significant market-share gains. They led us to believe that the number-four position—heretofore as remote to this company as Pluto is to the sun—was in fact an attainable, though risky, goal. And most important, they helped us find a credible way to demonstrate the incredibility of the insisted-upon goal.

In the end, the client received enormous value: It was steered away from a suicidal course of action based on reasons it could take to heart. And it was steered toward specific actions that should dramatically boost share and profitability. We also received the humbling lesson that only our client's stubborn insistence on an out-of-reach goal provoked really creative hypothesis generation.

■ ■ ■ ■ ■

I hope I've made it clear that these three tools—the executive in the parking lot, the focus group, and the stretch target—should carry a vibrating red warning label: "Use for idea creation purposes only. The hypotheses generated using these tools have not been tested in real business situations. Beware of groundless rapture and strategic intoxication."

The next step, according to OCE, is hypothesis testing and falsification.

Attitudes and Tools for Testing and Falsification

It is just as important for a company to have the right attitude toward the process of testing and falsification as it is for it to have the right attitude toward hypothesis generation. The task of creating the right attitude is more difficult here than in the hypothesis generation phase. It requires all your subtlety and skill as a manager because

- Somebody's ox, in the form of a pet hypothesis—possibly *yours*—is going to get gored.
- Conflict is inevitable.
- Ambiguity is rampant.

Whereas the proper attitude for hypothesis generation is to let joyous fancy fly free, as von Oech says, the proper attitude for testing and falsification is that of a stern and unyielding judge of the facts.[9] The necessity of testing accounts for the inevitability of the ox-goring and conflict. A moment's thought also reveals why it accounts for the rampant ambiguity. Testing and falsification, I argued in chapter 1, can rule out strategic hypotheses and narrow what *may* be right, but they can't prove that any strategic hypothesis *will* be right. In addition, there's always some room for argument about what the testing showed. That's not only human nature, but also the nature of thought. So at some point, the skilled manager, pressed for time and seeing that her team has reached the limits of its patience with conflict, cuts through the ambiguity and makes a decision.

So much for the attitude. The tools for testing and falsification can range from actual field trials of the strategy in selected markets against selected competitors to what I call *laboratory exercises.*

I limit the following discussion to laboratory exercises, because it's difficult to say anything general about strategic field trials. Suffice it to say that they suggest themselves almost immediately when the hypothesis is formulated: "An insurance policy aimed at state government workers! Let's try it in Des Moines!" Field trials may be expensive or cheap, depending on their size, the product, the service, and the cost to reach market. Since they often involve product development or brick and mortar or software development, they often irritate the CFO.

Field trials that require tossing a real, if prototype-stage, product or service into the marketplace exacerbate the ambiguity in any testing situation. They must be localized, inviting the charge that their success or failure is linked to (conceptual or geographical) locality. Then, too, budgets usually allow only limited time for trials and for only a few permutations of the product or service that is the core of the strategic hypothesis. It's hard, then, to distinguish a failure from a near success. The champion of the hypothesis can cry, "If only we'd continued a little longer. Or if we had just changed this or that."

Finally, these trials always seem to tip off the competition to your plans. Nevertheless it's hard not to like field trials. No paper exercise can replicate the in-your-face experience of the fate of ideas made tangible. The history of product development suggests that launching prototypes into the marketplace as *probes* that can be continually refined may be superior to relying on paper-and-pencil market research.[10]

The Laboratory as Courtroom

The philosopher Immanuel Kant was the first and best thinker to articulate the notion that the laboratory is a kind of courtroom. The scientist, he says, puts questions to nature in the form of experiments and sees how nature testifies. Both courtroom and laboratory are arenas for orderly and constructive conflict among ideas, as I suggested in chapter 1.

This laboratory, this courtroom, where the fate of hypotheses is determined, is furnished with judges: you, your colleagues, or your boss. It is furnished with two teams, each striving mightily to make the best of its case. And it is furnished with sources of evidence: witnesses and reports.

This isn't fantasy. There is no reason in the world why real companies facing real strategic decisions shouldn't formalize their strategic hypothesis testing and dignify the process with a place of its own. The key elements that are novel in this view are the teams in ritual conflict and the fact-based reports.

The Teams

It's just like *Perry Mason,* only the people are smarter, the acting is worse, and the outcome can't be predicted. If you are using consultants, you have by now heard their grand strategic vision. As I suggested earlier in this chapter, that should not be the end of the consultants' involvement. Their vision corresponds to the particulars of an indictment in a court of law, including all the compelling reasons why there is cause to believe that their strategic vision is true.

Consider your consultants as *counsel for the plaintiffs.* They make the case for a change in your company's current strategy. But here's the twist: I suggest that you hire *another* group of consultants to serve as counsel for the *defense* or designate a top team of employees to do the job. And what is that job? First, to attempt to falsify the hypothesis of the plaintiff, using the tools of logic and fact gathering. Rhetoric may be left at the door.

Second—and here's where we diverge from the American judicial system and veer toward the Napoleonic code—it's not enough to pick holes in what the plaintiffs are presenting. After all, we're trying to come to a decision about *what to do.* The defense team must come up with a compelling *alternative* strategic case. In a world of explosive change, doing nothing—which would be the outcome of a defense that falsified the plaintiff's case but offered no substitute—would be as much of a strategic direction as making any change the consultants might suggest. And the defense needs just as much energy, if not more, devoted to its falsification. This is the real shoot-out—not the beauty contest held when you're first choosing a consulting firm. And by the way, the usual rules of courtroom evidence and discovery prevail. The defense gets to have a good look at the plaintiff's fact base and logic, and vice versa.

Third, the team that *loses* the case doesn't suffer monetarily or in career progression. We can draw a lesson from the way Sony handles people when they present the results of their research and development in these shoot-outs:

> We constantly have several alternative projects going. Before the competition is over, before there is a complete loss, we try to smell the potential outcome and begin to prepare for that result as early as possible. Even after we have consensus, we may wait for several months to give the others a chance. Then we begin to give important jobs [on other programs] to members of the losing groups. If your team doesn't win, you may still be evaluated as performing well. Such people have often received my [Sony's top R&D manager's] "crystal award" for outstanding work.[11]

Ideally, the strategy shoot-out shouldn't ultimately be a matter of winning or losing. Instead, a savvy judge will see the end of the plaintiff and defense presentations as an opportunity for a synthetic moment. The judge (or judges) should

- Take what's best from both presentations and build a strategic synthesis.
- Focus on irreconcilable conclusions and call for tightly defined fact-finding to decide the issue.
- Spot the true believers on each side and put them in charge of strategy implementation. This is sometimes a very good thing to do with a *sore loser.*

The U.S. Army has taken the idea of using opposing teams to perform hypothesis testing to a fruitful extreme, while institutionalizing a way to synthesize the results. At its 1,000-square-mile National Training Center, it fields an *opposition force.* Units throughout the army are rotated in for mock combat, and the opposition force "usually wins."[12]

It's interesting that within five years of the first of these trials by (mock) combat, the army realized that they bore far more utility than mere training exercises. In fact, the army discovered that these exercises could be used to test war-fighting doctrine.

> At first, each unit coming here [to the National Training Center] tended to learn the same lessons as the one before it. The Army kept repeating a major mistake of Vietnam—where officer-rotation policies, it was said, meant the U.S. didn't fight for nine years but fought for one year nine times. To avoid that trap, it invented the lessons-learned process and, in 1985 founded the Center for Army Lessons Learned [CALL] at Fort Leavenworth, Kansas.[13]

CALL distills and codifies what the mock combat reveals, as well as experience in not-so-mock deployments such as those in Haiti and Bosnia. It sends out interviewers to talk to front-line soldiers and local authorities. It reviews raw information such as intelligence estimates and after-action reports. Professor David Garvin of Harvard Business School wrote a case study of the CALL process. The school has sold $289,000 worth of copies of the videotape of the Harvard study to businesses eager to learn how it was done.[14]

The army system includes three fundamental building blocks, as does our system of strategy on trial:

1. A competitive situation with opposing teams engaged in simulated struggle.
2. Independent referees and judges not only to determine who won but also to manage a process of uncovering.
3. The lessons that are to be learned and the next steps that need to be taken to improve the doctrine or strategy.

The Strategy Presentation

The presentation of strategic hypotheses consists of only four things:

1. *Assumptions:* The *givens* that both parties accept as so generally agreed upon that no one would bother disputing them. In the Kmart example from chapter 1, this might be that the company is losing share in key locations to Wal-Mart.

2. *The logic of the strategy:* This is the discussion of the relation between the *if* and *then* clauses of the proposed strategic hypothesis. The logic of the strategy specifies why people in their right minds should expect the proposed consequence, provided the conditions are true. In the Kmart–Wal-Mart example, it is the logic of the relationship between Kmart's store upgrading and expected increased market share.

3. *The fact-based reports that support the logic:* These are, for example, the quantified market research studies that the plaintiffs in the Kmart *case* might provide, the results of a pilot-store upgrade, or a set of comments from customer interviews.

4. *Refutation:* The refutation of the opposing team's logic or facts.

Fact-Based Reports: The Attractions of Discrete Choice Modeling

The fact-based reports play the role of exhibits and witnesses in our game of strategy on trial. There are all kinds of fact-based reports, ranging from teardown analyses of competitors' products to in-depth analyses of competitors' cost structure; from analysis of competitors' public statements to the results of private investigators' dumpster diving (analysis of a competitor's garbage).

The number and variety of market research techniques are boundless, but there is one that deserves more attention than it has received. It's called *discrete choice modeling* (DCM). DCM is a statistical technique that looks at buyers' actual choices in the marketplace. It examines actual purchase data (or data from simulated shopping experiences in your company's market research laboratory) and identifies the implicit trade-offs people make between prices and features, including the choice not to purchase at all. I think it deserves special attention because it

- Captures the consequences of nearly all factors affecting the revenue side of a strategy, including product features, distribution options, the value of a brand name, and even the value of proposed advertising messages.
- Reflects the purchasers' ability to choose not to buy or to postpone buying.
- Can be used retrospectively, based on actual behavior—for example, the implicit value purchasers put on various features of cell phones can be determined from the choices they actually make in the marketplace and the prices they pay.
- Reflects the trade-offs purchasers make every day among features, price, and availability, without relying on self-reports of motivations or intentions to purchase.
- Captures the effects of competitor reaction on revenue and market share along those same dimensions and recognizes the dynamic quality of competition.
- Identifies and distinguishes the rational and irrational components of marketplace decisions and quantifies both.
- Can provide the basis of an interactive computer model that allows your company to play competitive *war games* to tease out the results of possible strategic moves.
- Can be successfully adapted to non-Western cultures.

One of the most remarkable capabilities of DCM is its ability to underpin an interactive computer model. Plaintiff, defense team, or judge can vary the features, pricing, or advertising messages of a product or service, and the model will forecast the impact of the variation on revenues and market share. Figure 6-2 is a screen from one such model (the names and numbers have been disguised).

Figure 6-2. Example of interactive discrete choice model screen (data disguised).

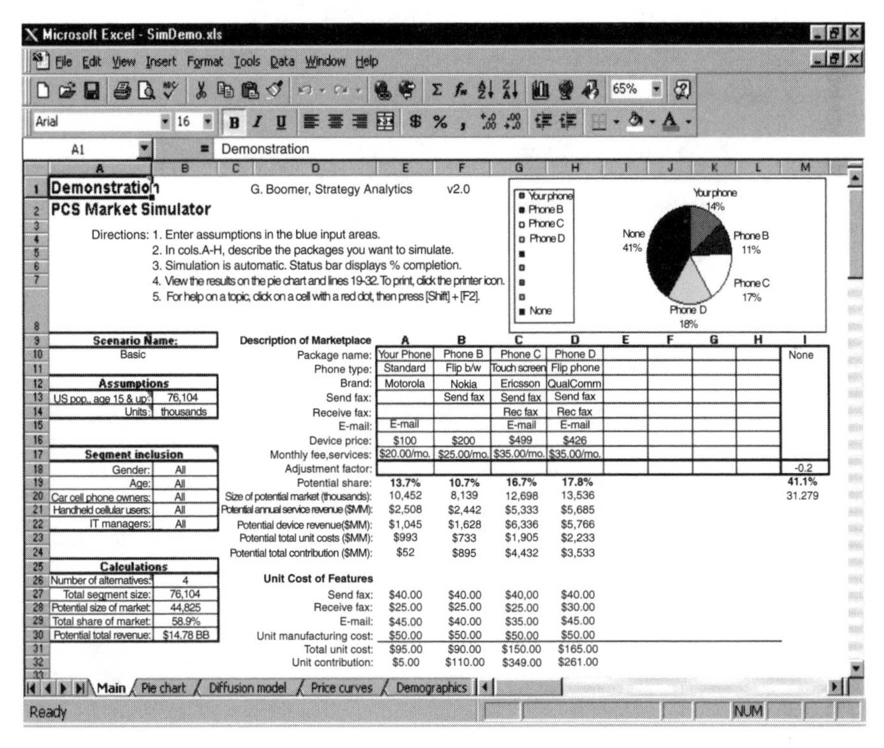

But more than one can play the interactive game. In fact, at the end of the presentations by the plaintiff and the defense, the judge can call as many people as she likes to play the roles of industry competitors. So, for example, you can see what might happen if you were in the cell phone business and decided to cut the price of one of your higher-end phones radically. This is illustrated in Figure 6-3, where QualComm has slashed the price of its (hypothetical) color-screen flip phone by more than half.

These powerful modeling capabilities provide an extra, welcome layer to your hypothesis testing and falsification efforts. Most market research reports provide, at best, a snapshot in time of market needs and preferences. The DCM model allows you to test a hypothesis that might pass with flying colors now but be falsified once competitors' predictable reactions are factored in. It is similar to the *"what if"* capabilities of a financial spreadsheet, but applied to customer preferences and behavior. (It's easy to see how the results of a DCM model could feed directly into a financial model, thus providing a superb integration of analytical results for the often warring marketing and finance departments.)

Figure 6-3. Results of changing pricing on cell phones.

Description of Marketplace	A	B	C	D
Package name	Phone A	Phone B	Phone C	Phone D
Phone type	Standard	Flip b&w	Touch screen	Flip color
Brand	Motorola	Nokia	Ericsson	QualComm
Send fax		Send fax	Send fax	Send fax
Receive fax			Receive fax	Receive fax
E-mail	E-mail		E-mail	E-mail
Device price	$100	$200	$499	**$200**
Monthly fee, services	$20/mo.	$25/mo.	$35/mo.	$35/mo.
Adjustment factor				
Potential share	**11.6%**	**9.1%**	**14.3%**	**30.2%**

DCM has its limits, too:

- It's not a modeling technique that is up to measuring industry-changing strategies, such as the acquisition of a competitor.
- It's not up to measuring the market response to the introduction of radically new products or services—for example, the introduction of inexpensive photocopiers during the mimeograph era, or the introduction of television.
- It requires fairly large sample sizes, typically several hundred, depending on how finely you need to cut the demographics.
- It's not useful for situations in which there are few buyers, such as the U.S. defense business, which has one domestic and a few significant foreign customers. It's more suited to markets of thousands and hundreds of thousands.

In other words, DCM has the limits of any other market research technique: It may be outstanding for hypothesis testing and falsification, but it is no substitute for strategic thinking. Once the *big idea* has crystallized, however, DCM can be useful in the attempt to falsify hypotheses about its market impact. For example, AT&T recently announced a deal to acquire TCI, the huge cable television company. It paid a handsome, above-market price, because it espied an opportunity to use TCI's infrastructure to enter the local telephone business and expand its broadband Internet service as well as its cable television business.

The value of this grand strategic move, in the end, depends on what consumers are willing to pay for that bundle of communications services. Do you see where this leads? The result is a fairly precise, falsifiable hypothesis about what consumers are willing to spend per month for various combinations of services, as well as how they feel about the domination of nearly all their communications by such a large global company.

This example shines a harsh light on the strengths and weaknesses of quantifying market research as a strategy falsifier. AT&T's strategic gambit could be falsified by needle-fine DCM research showing that the expected revenue wouldn't be worth the premium AT&T would have to pay for TCI. But no quantified market research is likely to uncover the potential for grassroots political opposition to the merger by those worried that the new company might bankrupt existing Internet Service Providers, which AT&T would naturally cut off from using its new, expensive network. Nor would one be likely to uncover the fact that this movement would be centered in Portland, Oregon, and Broward County, Florida, and gain the support of a federal district judge.[15] There simply is no royal road, except the road of thought.

■ ■ ■ ■ ■

Does this feel upside down to you? The process of strategic hypothesis development outlined here turns on its head the traditional way of thinking about strategy development. The typical view—the one you learned in high school as the *scientific method*—is that you start by collecting facts, then formulate a hypothesis, then verify the hypothesis with experiment.

The process for opportunity creation, discovery, and testing developed here is almost the precise reverse of this method. I suggest that you start by developing lots of ideas, distill the worthy ones into hypotheses that can be made precise, then attempt to falsify them by searching out those facts that make a difference to each strategic hypothesis. Along the way, we use the social judo of competition leverage and the human tendency to prove oneself right (at almost all costs) to get the testing and falsification job done.

Happily, what seems upside down is very much right side up. First, no thinking business executive is without thoughts about what works and what doesn't. He or she doesn't start with a mass of observations or facts but with insight, intuition, and a guess. Second, the best strategic consultants start out the same way. Their experience and company interviews give them a starting point for thinking about what the range of *right* answers might be. Then, with limited time and a big budget, they hunt down the facts that help them sort among the *live* range of hypotheses. (My quarrel with them is that they spike inconvenient facts, employ a narrow range of strategic ideas, and make little attempt to actually falsify their favorite hypothesis.)

In short, they do what really bright senior executives would do if they only had the time. And, interestingly, they do what the most productive experimental scientists do as well. Scientists start with a hunch or even a point of view based on training, a political agenda, or something else—you name it. Then hypotheses are formulated. And then experiments are designed. It's what real people with real budgets and burning ideas really do.

The literature on scientific method is extensive. Many philosophers and some historians of science have argued, in fact, that there are no facts without an embedded hypothesis: There are no raw sense data. We don't need to go that far to postulate that it is extremely uneconomical to go rampaging around this wide, diverse, confusing world for facts before having some idea of what we are looking for. Figure 6-4 illustrates these two views of hypothesis creation and testing. Which one accords with *your* experience?

To Do: Building a Strategic Dream Home

In some ways, the first steps toward building a corporate strategy are like those involved in building your dream home. If it's going to be yours, to live in and pay the mortgage on, you can't simply subcontract out to an architect or builder your requirements, feelings, and ideas. Your requirements list is a starting point, but there is more than one solution to that list. If you don't believe that, give the same list to two architects. You can get an architect to provide thought starters, tell you what's possible and impossible within your budget, uncover hidden conflicts in traffic patterns and structural requirements, and provide details on energy consumption, light, proportion, and texture. But no one in his or her right mind would spend big money on the *first solution* to pop into an architect's head without pushing back, testing, or seeing a model.

Figure 6-4. Two views of scientific discovery.

Textbook

1. Gather lots of facts
2. Look for patterns
3. Develop experimental plan to test pattern-based hypothesis
4. Perform experiment

Real World

1. Have hunch, insight, or intuition
2. Sharpen into one, two, or more precise, competing hypotheses
3. Develop experimental plan to falsify hypotheses
4. Perform experiments to gather pertinent observations that falsify, quantify hypotheses

So, the to do's from this chapter are straightforward:

1. First, a responsibility check: Has our company subcontracted responsibility for our strategic thinking, or are we prepared to do the heavy lifting ourselves? In either case:

- The plaintiff team is going to be ______________________.
- The defending team is going to be ______________________.
- The presiding judge is going to be ______________________.
- Our method for resolving irreconcilable differences at the end of the case is going to be ______________________.

2. Do we have methods in place for developing *lots* of ideas? In particular:

- Are we still looking for the *one right idea?* If so, who is doing the looking, and who isn't convinced that the best route is to have many ideas?
- What methods are we using to break free of the mental ruts we all fall into?
- What methods are we using to reward people who have lots of ideas? And to share them?
- In the hunt for a broad range of strategic hypotheses, are we using our executive talent well? Do they get out to the front lines in the marketplace?

- How are we going to collect and make a first-pass cut of the ideas selected?

3. Similarly, are we *asking enough* of methods such as focus groups? Conversely, are we relying too much on them as hypothesis testers rather than as thought starters?

4. What quantitatively based methods are most appropriate for attempting to falsify the plaintiff's key strategic hypothesis? The defending team's strategic hypothesis?

■ ■ ■ ■ ■

Now that tested, formulated strategic hypotheses are in hand, the next step is to get them out the corporate door.

Notes

1. Quoted in Roger von Oech, *A Kick in the Seat of the Pants* (New York: Perennial Library, Harper & Row, 1983), 24.
2. Ibid., 95. See also James Adams, *The Care and Feeding of Ideas* (Reading, MA: Addison-Wesley, 1986); Charles Thompson, *What a Great Idea!* (New York: Harper Perennial, 1992), 9–12; James Brian Quinn, "Managing Innovation: Controlled Chaos," *Harvard Business Review*, May–June 1985, 75.
3. James Billington, *Customer-Driven Innovation, Harvard Management Update* (Boston: Harvard Business School Publishing, 1998), 3–4.
4. David S. Landes, *The Poverty of Nations: Why Some Are So Rich and Some So Poor* (New York: W. W. Norton, 1998), 288–89.
5. Quinn, "Managing Innovation," 77.
6. Billington, *Customer-Driven Innovation*, 1.
7. Personal interviews, current and former GM and Japanese company executives.
8. George M. Prince, *The Practice of Creativity* (New York: Collier Books, Macmillan, 1970).
9. Von Oech, *A Kick in the Seat of the Pants*, 89–112.
10. Gary Lynn, Joseph Morone, and Albert Paulson, "Marketing and Discontinuous Innovation: The Probe and Learn Process," *California Management Review*, spring 1996, 8–37.
11. Quinn, "Managing Innovation," 78.
12. Thomas E. Ricks, "Army Devises System to Decide What Does, and Does Not Work," *Wall Street Journal*, 23 May 1997, A1, A10.
13. Ibid., A10.
14. Ibid., A1.
15. Kathy Chen, "Another Vote to Open up Cable Lines Means More Complications for AT&T," *Wall Street Journal*, 14 July 1999, B6.

Chapter Seven

Strategic Breakthrough and Exploitation: Making the Right Choices and Choosing the Right Tools

The only security is in the pursuit of opportunity.
—DOUGLAS MACARTHUR[1]

"Well begun is half done," says the proverb. In the last two chapters, I discussed how to create, capture, and test good strategic beginnings in the OCE cycle. But that is only the beginning. The next phase of the strategic cycle is building something out of that wonderful beginning: the development of strategic *breakthroughs* and their successful *exploitation.*

In chapter 5, I defined a *breakthrough* as "the production of the first measurable success from the strategic hypothesis." Likewise, *exploitation* is "doing what it takes to enlarge on the success of a strategy." I refrain from calling these phases parts of *every* strategy, because they don't come packaged automatically with the grand strategic hypothesis. Rather, I choose the phrases *production* and *doing what it takes to enlarge* to emphasize that unless you are very lucky, even the cleverest strategic idea does not by itself entail success. Breakthroughs and the rewards of exploitation are not the inalienable birthright of a great strategic hypothesis.

You might say that compensating for human intellectual frailty was the objective in the previous chapter. That's why we put strategic hypotheses on trial, complete with plaintiffs, defense teams, and exhibits. Compensating for the difficulties every organization has converting ideas to action is the objective of this chapter. This chapter is about

- The crucial *choices* top management must make about breakthroughs and exploitation
- How to *direct resources* to take advantage of exploitation opportunities.

Choices for Breakthroughs

The crucial step in producing a breakthrough based on a strategy is choosing what kind of *real-world realization* the surviving strategic hypothesis (or hypotheses) should have. The real-world realization is the concrete form a hypothesis takes. In chapter 1, for example, we talked about a hypothetical Kmart strategy including larger, renovated stores with higher-scale merchandise, higher lighting levels, and low everyday prices. The real-world realization of that strategy would be the specification of exact locations and sizes of the stores, their layout, merchandise selection, managers, compensation schedules, training requirements, and the thousand other details that translate the strategic hypothesis not simply into time-honored action plans but into the nitty-gritty of architectural blueprints, recruiting advertisements, and orders to vendors. Up to this point, of course, a strategic hypothesis is nothing more than words—important, change-laden, career-risking words, but no more than that.

The crucial options are as follows:

1. *Probe with pinpricks:* Use numerous but separate low-cost efforts to determine the best real-world realization of the hypothesis.
2. *Drive and concentrate:* Focus all effort on a single, complete real-world realization of the hypothesis.

The idea of the first alternative is to develop a portfolio of real-world realizations of the strategy and see which one works best. In the Kmart example, it would mean a portfolio of store designs, merchandise selections, and so forth. This approach gets the strategy rolling, but it also extends the concept of testing the hypothesis, refining it, and getting more precise about it all the time. It is the strategic version of the Japanese concept of *kaizen*—continual improvement on a core vision.

The idea of the second alternative is to commit all resources to a well-thought-out, extremely complete real-world realization of the strategic idea. Once the concept of larger, well-lit stores with higher-scale merchandise is selected, this alternative typically calls for squadrons of experts and task forces to determine *the* optimal merchandise mix, *the* optimal lighting level, *the* optimal store size, *the* optimal manager compensation package, and *the* optimal store locations. The two alternatives are illustrated in Figure 7-1.

Pinprick Examples and Models

The drive-and-concentrate alternative derives easily from the exhaustively examined, tested, and revised strategic hypothesis, and it is certainly most companies' default mode of operation. It's what you see in most action plans appended to strategic recommendations. That's why I want to spend more time discussing *probing with pinpricks.*

Microsoft is an example of a company comfortable with probing with pinpricks. It also has the cash to do it in style. In 1999, it was generating free cash of about $9 billion per year, and despite a $5 billion investment in AT&T, it still had $18 billion in cash.[2] It realizes that no one knows how the Internet will develop, or whether or how telephony and entertainment will converge radically with personal computing (or relegate it to obsolescence). So it's not concentrating all its wealth on a single strategy. Instead, it has invested or created alliances in cable television, set-top Internet boxes, television broadcast-

Figure 7-1. There are two methods of seeking strategic breakthroughs.

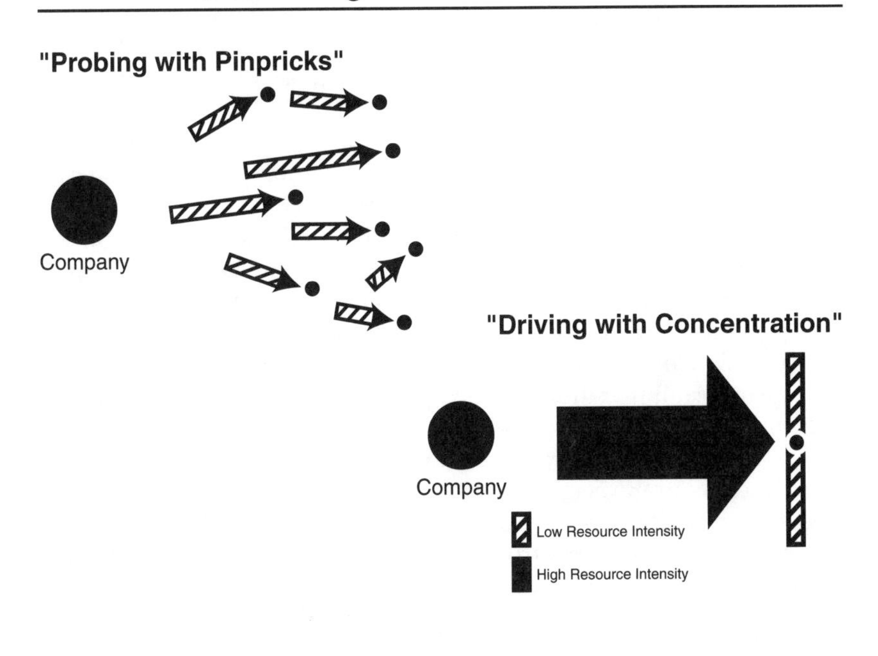

ing, satellite communications, Internet service, long-distance telephone companies, and multimedia software. A lot of these ventures aren't making any money; some are long-term investments generating no immediate cash. Nearly all of them not only make Microsoft less dependent on the fate of the personal computer but also prepare it to take the pole position in the future, whatever that turns out to be. It must be nice.

Charles Schwab & Co. is another example of probing with pinpricks. From 1982 to 1996 it tried no fewer than six different times using six different technologies to bring electronic stock quotations to thousands of retail customers. The first five attempts were what Charles Schwab calls "noble failures." The sixth attempt, Schwab's 1996 Internet offering, has been a resounding success—a success paid for by Schwab's persistent use of the pinprick approach.[3]

Thermo Electron is far smaller than Microsoft, with only $3.9 billion in 1998 revenues, compared with the Redmond giant's $14.5 billion.[4] Although Thermo Electron doesn't have the cash of Microsoft, it's found its own way to field a portfolio of pinpricks. Thermo Electron became famous in the 1980s and 1990s for spinning off its bright ideas into public companies in which the parent owned a stake. The result has been a wide range of products, strategies, and services, from mammography equipment to satellite communications equipment to propane-powered road lighting. Sometimes these so-called *public subsidiaries* spin off their own public subsidiaries. For example, ThermoTrex, a $343 million public subsidiary, has a number of product lines and three smaller subsidiaries. One of these is ThermoLase, which developed laser-based hair removal and then bought a network of expensive health spas to ensure a source of customers for its hair-removal services.

Thermo Electron has found a way to field a wide diversity of strategic initiatives while limiting risk to the parent company—raising capital for these initiatives in the public markets. Sometimes this innovative but fundamentally conservative approach is right on the mark; sometimes not. ThermoLase recently announced the sell-off of its spa centers: a pinprick that didn't find its target. But because the pin was cushioned by its two-tier public subsidiary relationship to parent Thermo Electron via ThermoTrex, the parent hardly felt the prick.

As a final illustration of probing pinpricks, we have a sketch from the world of product development. A fascinating study describes the adventures of Searle Corporation (Monsanto, today) with its artificial sweetener, NutraSweet (aspartame).[5] In delightful accord with what I've said about the role of serendipity in discovery, a Searle scientist searching for an ulcer treatment discovered aspartame in 1965.[6] Searle attempted to use NutraSweet to break into the sweetener market in the early 1970s by probing the markets in chewing gum, ice cream, carbonated beverages, powdered drinks, breakfast cereals, whipped toppings, and spoon-for-spoon, tablet, and packet sugar replacements. It started with high hopes, and efforts directed toward the cereal, carbonated beverage, and spoon-for-spoon markets.[7] The cereal probe foundered on the fact that when General Foods removed the sugar and substituted the far-sweeter-per-ounce aspartame, the cereal lost much of its bulk. Technical and Food and Drug Administration approval difficulties hampered the rest until 1982. Searle was so discouraged that it wanted to drop the product and sell its rights to NutraSweet. But it couldn't even admit defeat and withdraw: It couldn't find a buyer.

Of all the probes, it turned out that NutraSweet's breakthrough success came in the form of sugar-substitute packets (the well-known brand name *Equal*). But even here there was an unexpected twist: Instead of taking share from saccharine (infamous for its bitter aftertaste), as Searle had expected, it took share from—you guessed it—*sugar*, an "entirely unexpected market segment." Sales exploded from $74 million in 1982 to $336 million in 1983.[8]

Parallels in Nature

Companies that probe with pinpricks follow the procreative path of insects, trees, and fish. To reproduce, these living organisms typically create hundreds or thousands of eggs or seeds. Only a tiny percentage are fertilized, fewer still survive infancy, and even fewer survive to maturity. One naturalist estimates that even under favorable conditions, fewer than one in a thousand pinecones becomes a pine tree.[9]

Contrast this approach to species survival with that exhibited by many mammals and some birds, which produce only one, two, or three offspring per year. Whereas adult insects, trees, and fish put little or no effort into making sure that their offspring survive—trusting to numbers and to built-in survival mechanisms—these other animals frequently invest tremendous energy in their few offspring. With so few offspring, they have a lot to lose. I recently viewed an osprey nest overlooking Lower Yellowstone Falls in Wyoming. It has maintained its precarious perch overlooking the roaring Yellowstone River canyon for over ten years. The female osprey lays but a single egg each year. This spring, the egg rolled out of the nest. That's it until next year. Among humans, this analog to the drive-and-concentrate approach has reached an apparent zenith, with parents tending their children for a decade and a half to two decades (and sometimes four!).

Probing Pinpricks in Military Strategy

The U.S. Army, in its pursuit of methods to destroy hypothetical enemies, makes a distinction that is exactly analogous to that between probing pinpricks and drive and concentrate. It's particularly instructive because the army is perforce a competition-oriented organization.

Robert Leonhard contrasts two of the army's attack doctrines. The one analogous to our pinpricks is the inelegantly named *recon-pull* (where *recon* is short for *reconnaissance*).[10] The other, analogous to our drive and concentrate, goes under the rubric *command push.* Leonhard distinguishes them as follows:

> A subset of directive control—an attack variation—is the idea of "recon-pull." . . . [L]eaving the initiative to subordinates, recon-pull allows subordinate commanders to move along different routes toward the enemy. Hopefully, one or more of the units will find an undefended or lightly defended "gap" that leads to the enemy rear. Seeking to exploit this opportunity to attack enemy weakness rather than strength, the whole unit will then respond to the forward subordinate commander's report, and all will move to penetrate the gap. In this way the friendly force refrains from "leaning into the enemy's strength," where even victory would be costly.[11]

The main features of the pinprick approach for creating breakthroughs can be distilled from Leonhard's dispassionate prose:

1. There is a *multiplicity* of separate efforts.
2. There is an *economical use of resources* in each effort (the implication of the phrase *"subordinate commanders"*).
3. There is an *attitude of humility*—that the exact answer to implementing the overall strategy isn't known until actual trials are made.
4. There is *no wholesale commitment* until the results of real-world trials are in.
5. No individual trial, by itself, is expected to complete a breakthrough; *nor is too much to be risked* in any one effort through persistence in the absence of concrete progress (the implication of the word *reconnaissance*).

The opposing approach, command push, is based on a command style that is called in German *Befehlstaktik* (*Befehl* means *"orders"*—in this case, detailed orders from on high). Leonhard describes it thus:

> The essence of this more traditional form of control is that the commander issues detailed orders, leaving subordinate commanders little freedom to deviate. Rather, the smaller units all move and fight according to the plan. Under this method, the key to victory is not to exploit opportunity, but to impose the commander's will upon the enemy with irresistible momentum through unified action.
>
> Having already decided where the attack will occur, the commander masses his teams, fire support and logistics assets and his recon assets along the axis of advance.[12]

This sketch manifests the crucial elements of what I call *drive and concentrate:*

1. The strategy, once decided upon, *gives voice directly* to its real-world realization, without further probing, testing, or other ado.
2. The strategist *commits all available resources* to the plan, imposing on the world (the market, the competition) through a high expenditure of resources, even if initial gains are few. ("They're not buying? Then fire the ad agency and double the budget.")
3. The approach *relies on intense, detailed planning and coordination of resources*—a complete action plan.

Leonhard later makes an interesting observation parallel to the one I've made about the near dominance of drive and concentrate (or command push) over probing pinpricks. His personal experience as an army officer is that official doctrine is recon-pull, but real-world action is command push. If we pause a moment to reflect on that, it's not surprising. In the army, as in business, it's very hard to give up a highly probable but more costly victory over which the commander has some control for a more speculative solution that is at the moment shapeless, unknown, and utterly dependent on the talent of subordinates, the commander's trust in them, their trust in one another, and a subordinate officer's natural unwillingness to drop his or her leading role to become a supporting player in someone else's show.[13]

Being Conservative or Aggressive with Breakthrough Risk

All things being equal, which they never are, the drive-and-concentrate alternative is far more risky than probing with pinpricks. It supposes that once we're done with the strategic hypothesis creation, we can discover or create a priori *the one best way* to realize the strategy in the real world. The probing alternative is more conservative and limits any given loss, while accepting the reality that no one knows how the future will turn out for any diversification concept.

So if the idea of a portfolio of pinpricks can have such appeal, why does experience suggest that most of the time executives choose to drive and concentrate? The reasons are both reasonable and several. And they are worth bringing to your company's attention, so that if your company chooses the drive-and-concentrate alternative, at least it does so for sound reasons. I sympathize with executives who, under pressure from their boards, investor pressure from Wall Street, and time pressure from their own timetables, decide that they must take one great shot at realizing the new strategy, hoping to fix any flaws later. Also, it's entirely understandable that a company may simply not have the depth of talent, creativity, and management to undertake more than one groundbreaking project at a time. It's also the case for many public companies that investor—read Wall Street analyst—comfort is tied up with being able to comprehend what a company's strategy is. The geniuses on Wall Street are notoriously impatient with experimentation. Further, many executives feel, with the justification of bitter experience, that making a root-and-branch commitment to a single, well-defined strategy—even if it's somewhat off the mark—produces better odds of success than refining a strategic hypothesis.

Then, too, some strategies are by nature *one-bet* strategies. For example, trying out a portfolio of variations on a strategy based on high unit volumes, economies of scale, and low prices is often just too expensive. How many enormous plants can a company afford to build? In contrast, the key components of a strategy can often be simulated before the company commits to scale-up.

Many executives despair at the resources tied up in creating pinprick probes. They think in terms of the effort it would take to create, say, a portfolio of prototype products (or, in our Kmart example, store layouts and merchandising sources) in order to field a number of real-world realizations of a strategy. But this is easily exaggerated: Each alternative should absorb about the same amount of total resources—namely, everything you can allocate to making a strategy successful. It will, after all, be the Hammer of your new overall company strategy.

.But there are often low-cost methods available for launching probing pinpricks. Whereas full working prototypes may be expensive, mock-ups or demonstration models of a product crucial to a strategy are less expensive. A press release announcing the target launch of a product or service and accepting *preorders* can be a pinprick. So can demonstration projects funded fully or partly by target customers. Multiple Web sites with different offerings, messages, and pricing can probe market and competitive response; successful ones can be developed further, converting a breakthrough into an exploitation.

Not to be overlooked are the approaches that already exist within your company. On a recent strategic consulting assignment for a large transportation services company, I had the happy experience of uncovering several real-world realizations of strategic approaches the CEO wasn't fully aware of (and there was no reason he should have been; we were dealing with a profitable but tangential business opportunity). The headquarters of this company was in a large southern U.S. city. Local service outlets were typically company owned, but not always. One money-losing Rocky Mountain outlet had been sold to an entrepreneur who was forging success out of failure. In another case, a subsidiary of the company had no company-owned outlets at all, but it implemented a novel management system to ensure that its leased outlets were productive. In a third case, a company-owned outlet in another southern city had created mold-breaking contracts with local customers. The only expense involved in turning these cases into real-world realizations for the agreed-upon strategy was the cost of learning what made them successful, how, and why.

Breakthrough Alternatives and Inherently Variable Environments

Finally, the choice of pinprick probes or concentrated drive should be grounded in the fundamental nature of your industry. Recall Sutherland's buckets of deterministic, moderately variable, severely variable, and indeterminate situations. A driving, concentrated approach to creating a strategic breakthrough is well suited to a deterministic industry environment. The more variable the environment, the more appropriate the probing pinprick approach.

But don't stop at characterizing the *prevailing* variability of the nature of the markets or industries your strategy is targeting. Searle launched probes into industry environments that leaned toward the deterministic and moderately variable—especially food processing. But as the Searle example suggests, we need to consider how radical a strategy is relative to that industry, and whether it will push the industry into a different category of variability. What Searle was proposing was radical, especially for cereal makers and carbonated drink producers. For them, it meant a whole new orientation, involving high risk and the conversion of a moderately variable environment to a severely variable one, as they moved toward launching sugar-free products.

Choosing between These Breakthrough Alternatives

What's important about choosing between these alternatives is that you:

- Make these choices consciously and rationally, or they will happen by default. Remember that a strategic hypothesis, even with all the detailed planning in the world, is just words.
- Avoid choosing a muddled middle way—halfheartedly backing one approach (to avoid betting the farm on resources and reputation) and losing the option of taking the other.
- Align your choice with the present and prospective variability of your strategy's target industry environment.

Crucial Choices for Strategic Exploitation

Breakthroughs are wonderful and exhilarating, especially if they follow on the heels of a well-made plan. They're the first reward for all that thinking, shrewd guessing, and hard implementation work. It seems to be in the nature of things, however, that there is always a surprise in store when strategic theory meets market and competitive reality. So let's turn our attention now to the next phase of strategy implementation, exploitation. Recall that by *exploitation* we mean doing what it takes to enlarge on the success of a breakthrough. Thus, in our hypothetical Kmart case, the first dozen stores under the new format are achieving the hoped-for market-share gains. Or at Searle, we are at the point where Equal, the sugar-substitute brand, is showing market acceptance.

Since the next step is to thrust all available resources into the strategy's new Hammer, it's prudent to understand why a breakthrough is a success and how the particulars shed light on the initial strategic hypothesis. Then we must choose between two important modes of exploitation: *in depth* or *sideways*.

Understanding the Reason for a Successful Breakthrough

Suppose that you run a twenty-unit hospital chain. Your strategic work suggests that it may be worthwhile to take advantage of the upsurge in consumer interest in wellness. Heretofore, like all hospital-chain executives, you've been more interested in filling beds. You get a report on the daily census, and you know that you are probably in the black when over 55 percent of your beds are filled. But you also know that there's an overcapacity problem in U.S. hospitals: too many beds. And there is a rising interest among patients in wellness.

This situation recently impelled the chain's board to undertake a complete strategic review.

As a result, you found that a large segment of the population was willing to pay something for wellness—willing to pay for the prospect of not getting sick. So you launched pinprick probes in each of your twenty markets. Each hospital unit administrator promoted some version of a wellness program. Today, you find what looks like a breakthrough at one of those units. One hospital has had success in a local market by promoting a wellness program in a lecture series, followed by a four-month-long series of health workshops led by high-profile physicians. Smaller and medium-size local companies that have performed cholesterol, blood pressure, and diabetes screening for their employees seem willing to appropriate up to $150 of the $299 fee for the participation of each of their highest-risk workers in the workshops. Meantime, this same hospital's new cardiac care center has not been a success—in fact, it's been bleeding cash, even though the local administrator has recruited two top cardiac surgeons to make it a surgery center of choice.

The first step is to understand the meaning behind this hospital's success. Success, as they say, has many fathers. In this case, it could be one or all of the elements in the real-world realization of the new strategy:

- The drawing power of the community's high-profile physicians. Perhaps the doctors see the program as a way to increase their exposure in the community at a time of static or declining physician income.
- The advertising message. Did it just happen to be on target?
- The interest shown by small and medium-size employers but not large companies and not individuals. Why is that?
- The price. The truth is, the hospital administrator pulled the $299 figure out of a hat—it was enough to cover expenses, with a little left over. It seems that the companies are happy subsidizing half and the employees are willing to spring for the other half.
- The demographics of the people attracted to the program.
- The time of program launch. The program was launched in early spring—just by chance. Were people interested in health in view of New Year's resolutions or the impending summer?

You can see how considering these factors feeds back into refining the strategic hypothesis. That helps make the strategic hypothesis *both* more precise—in terms of what a successful offer to the market will be—*and* more generalizable; knowing why this breakthrough worked provides confidence for the other nineteen hospitals.

Exploiting this success might mean customizing wellness program approaches for different local communities (e.g., black, white, Hispanic) and then offering the programs in surrounding neighborhoods rather than in hospital meeting rooms. It might mean advertising the service on the radio using the voices of local physicians. It might mean promoting the program in venues where managers of small and medium-size companies congregate—the Rotary Club, for example.

If this sounds like common sense, be reminded that exploiting the breakthrough could mean wiping out the budget for the cardiac center and sending cardiac patients to competitors, then using the freed-up funds to push the wellness program. Although the cardiac center in this hospital may be losing buckets of money, doing this would be terribly difficult. Even failures have organizational constituencies that push for their survival. But a new Hammer must sometimes be starved to provide expansion resources. Imagine the screams from the cardiac center staff!

Succeeding in the exploitation phase requires making a stark choice: focusing on the few items that make the breakthrough what it is, or shifting resources *sideways*. Let me explain the difference by illustration.

Compaq: Example of Sideways Exploitation

In the 1980s, Compaq was both the first highly successful IBM PC clone builder and the highly successful creator of luggable carry-on-size personal computers that were hauled by accountants and consultants with love, dismay, and aching backs all over the continent. Many people forget that these machines first brought Compaq into its eminent market position—a far cry from the full line of desktops, notebooks, and servers it produces today. Compaq's core capability at that time was the ability to clone IBM's BIOS technology and make a truly IBM-compatible clone. That, plus clever design, made the PC possible.

Soon after, Compaq exploited its clone capabilities and turned the company into a leading desktop clone manufacturer. It placed a crucial bet when it sought to bypass the Intel 286 market quickly and calculated that its technological prowess and strong brand name could carry it into the new 386 technology as a marketer of desktop computers. By expanding its presence in the clone marketplace, Compaq was choosing a sideways exploitation—tackling markets and competitors (other desktop PC makers) adjacent to its transportable market niche.

Theoretically, Compaq could have chosen to exploit its breakthrough by moving further into the transportable market—concentrating on distributing the product, making it lighter and more powerful, developing variations for particular customer segments (e.g., extremely rugged versions), pushing unit costs lower, and so forth. And another company has actually done this: Toshiba is the leading volume, technology, and design player in laptop computers, leaving the desktop to others.

Compaq, however, chose a riskier but potentially more rewarding exploitation option. I say riskier because there were already a number of clone makers emerging, with greater or lesser degrees of *true* IBM compatibility. In 1997 and 1998, Compaq used its then-dominant position as a corporate supplier of PCs to expand into the flanking corporate markets of network servers and, with its purchase (announced in July 1997) of Tandem Computers, Inc., into mainframe computers. Then in 1998, Compaq announced the acquisition of Digital Equipment Corporation (DEC), moving sideways again into minicomputer and chip design. No doubt future Tandem, DEC, and Compaq computers will be designed to link especially well with one another, creating an estimable competitive advantage for Compaq in serving large corporate customers.

Sideways exploitation moves a company inexorably away from its core breakthrough. While Compaq was pivoting on its PC business to snap up Tandem and DEC, it suddenly found that very Pivot under attack, especially by direct sellers Dell and Gateway. Dell in particular found a build-to-order, direct-sale, high-inventory-turnover strategy that started savaging Compaq in its core business market. In 1999, Compaq chief Eckhard Pfeiffer, who had orchestrated the sideways moves, was forced out. Now founding venture capitalist Ben Rosen is refurbishing Compaq's Pivot by focusing on turning around the PC business: focusing on four, not forty, distributors; selling PCs on the Web; and completing Pfeiffer's cost cutting.

UPS: Example of In-Depth and then Sideways Exploitation

For most of its history, United Parcel Service has been a paradigm of exploitation in depth. It focuses on wringing the last drop of efficiency out of its system. It dominates the parcel delivery market through the relentless, Tayloristic pursuit of logistical efficiency: driver scheduling and routing, package collection, and sorting. Drivers are trained how to get into and out of trucks, how to shift gears, when to leave the truck idling, when to risk a parking ticket. Operations run with military precision. UPS trucks are washed and cleaned every day, promoting a longer life span than competitors' trucks. The result is that no one can touch UPS for small parcel delivery nationwide. It remains highly profitable, even with a high-cost, unionized workforce. (Workers and management together own the company, although it announced plans in mid-1999 to sell a portion to the public so that it could establish the open-market value of its shares for use in acquisitions.)

UPS has achieved this profitability and market reach through absolute dedication to efficiency and cost control, even at the expense of customer service and satisfaction. For this *excellent* company, the customer is not number one. *Efficiency* is number one. The UPS strategic hypothesis encompassed the bet that wide service area and low cost would trump customer service and flexibility in the marketplace. Indeed, UPS is notoriously inflexible. Drivers don't wait for packages: Cutoff times for pickups are rigorously observed—*enforced* might be a better word. The pursuit of speed means that no extra time can be expended caring for fragile packages. One UPS driver told me at my loading dock: "If it can be broken or damaged, please don't ship with us." An eighty-pound weight limit is also strictly observed; forget it if you're a few pounds over. Pricing schedules are marketed on a *take-it-or-leave-it* basis.

For the longest time, there was no tracing of ground-delivered packages at all; then in the early 1990s, tracing was introduced with premium three-day ground service. Any tracking had to be performed from the origin side. If the expected package didn't arrive, the intended recipient had to contact the shipper to get any tracing action launched.

This ruthless attention to efficiency led UPS to stretch to serve even the remotest rural communities. It could then justifiably claim to be the only truly national ground delivery service capable of reaching every address.[14] In the foothills of the Bighorn Mountains of Wyoming, a UPS truck will pick up and deliver to the remote office of a former Girl Scout camp located on a dirt road half an hour from the nearest town of Ten Sleep. UPS must do it profitably, too; it has the advantage of being the only nonpostal parcel delivery service available there.

In the mid-1980s, UPS realized that this route of exploitation in depth was about played out. Its national ground network was complete. I suspect that it was becoming harder to extract significant cost savings from time-and-motion studies. And UPS's position as the gorilla of U.S. package delivery was unchallenged. Athough local package delivery services might *cherry-pick* specific high-traffic corridors (e.g., San Francisco–Los Angeles or New York–Boston) for business traffic, only UPS and the U.S. Postal Service served the entire country, both business and residential.

Meanwhile, the creation of its air express service, based on dedicated aircraft routed through a hub system, had created a whole new market for shipping documents, an item that had rarely been entrusted to Big Brown. And there was huge growth in catalog company and direct marketing shipments—long a staple customer base for UPS. Catalog companies first offered two-day and overnight air express as an extra-cost alternative to last-minute shoppers. Now for some companies, Land's End, for example, two-day air shipment is the standard service.

In another move sideways, UPS entered the overseas ground delivery and air express markets, which promise more growth than domestic markets. They don't want to cede this to Federal Express, which took the plunge into Asia in the 1980s (and initially lost huge amounts of money) with the purchase of Flying Tigers, or to such international competitors as DHL and TNT.

The sideways move of UPS into domestic air express could be based on a very strong Pivot, precisely because it had exploited that Pivot in depth. Operations analysis revealed that with some tweaking, UPS could use its existing ground delivery system (those ubiquitous brown trucks and vans), its broad network of sorting and distribution centers, and its huge vehicle fleet for both ground and air express services by modifying the system's operations. In other words, trucks, sorting centers, and pickup counters could handle both time-sensitive, high-reliability air express packages and time-indefinite, *adequate-reliability* ground packages. It didn't need to create a dedicated ground fleet as did Federal Express or contract with local cartage agents as did Airborne Express or Burlington Northern Air Freight (now BAX Global). And its Pivot wasn't just operational and asset based. UPS had decades of experience in implementing efficiency: It had a kind of *knowledge* Pivot.

Today, UPS has finally begun adding competitive *customer service* features. The reluctance is almost palpable. But its tracing and tracking are now state of the art, including Web-based customer tracking even for lowly ground packages. And its reliability has climbed from mediocre to an acceptable par with its air express competitors.

Exploitation's Master Rule

Exploitation, whether it is *expanding sideways* or *going deeper,* adheres to the rule of concentrating resources on *adjacent opportunities.* Thus, for Compaq, business desktop computers were adjacent to its business transportables; servers were adjacent to business desktops. What we don't usually see are successful leaps to nonadjacent businesses—for example, if Compaq had leaped immediately from business transportable to entertainment-oriented home computers without the intervening stage of selling business-oriented personal computers, then home-oriented computers, then finally entertainment-oriented computers.

UPS, for most of its history, has gone *deeper.* Its adjacent opportunity was to extend itself geographically—not immediately to the remotest communities and rural homes, but further and further. And

UPS got really good at it. Only then did it exploit other adjacent opportunities by adding expedited three-day ground service, then overnight and two-day air express, overseas express service, and even same-day shipment services.

Exploitation Choices and Inherently Variable Environments

Perhaps the exploitation choices UPS and Compaq made reflect the personalities of their CEOs or their company cultures. But before we ascribe the choices to such imponderables, it's worth noting that Compaq faced a far more variable environment than UPS. In a fast-moving, variegated, highly variable environment, sideways exploitation can make sense. Reason: It seizes opportunities that are current rather than building strength in the area of breakthrough success, which may change tomorrow and where the causal link between the investment made and the results desired is weak. The computer industry moves quickly and sometimes, it seems, by stealthy steps—few foresaw that the Internet transfer protocol would become a consumer good. By contrast, the *P* in UPS might have stood for "phlegmatic" when it comes to industry change (until 1970s deregulation boosted industry variability). UPS had every reason to believe that getting better and better at the one thing it did would result in the strategic outcomes it had targeted. Figure 7-2 is a summary of how exploitation choices align with environment variability, and Figure 7-3 summarizes your breakthrough and exploitation choices.

■ ■ ■ ■ ■

Figure 7-2. Exploitation choices typically conform to variability of environment.

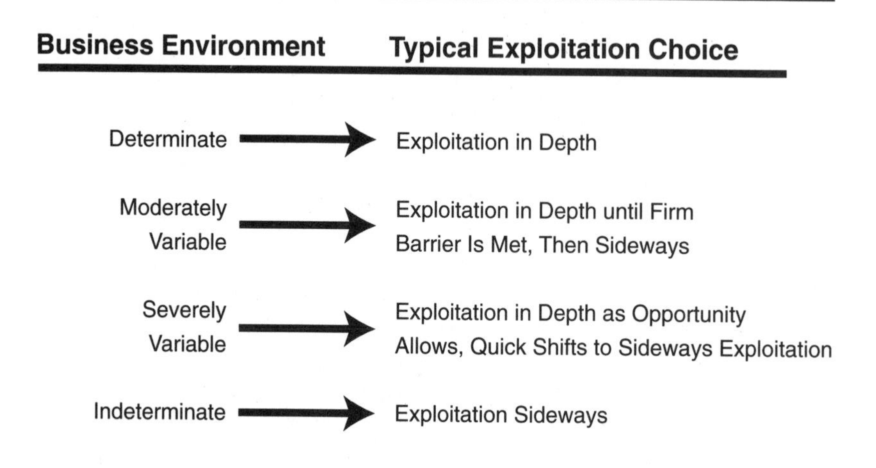

Figure 7-3. Two exploitation choices face the strategist after the breakthrough.

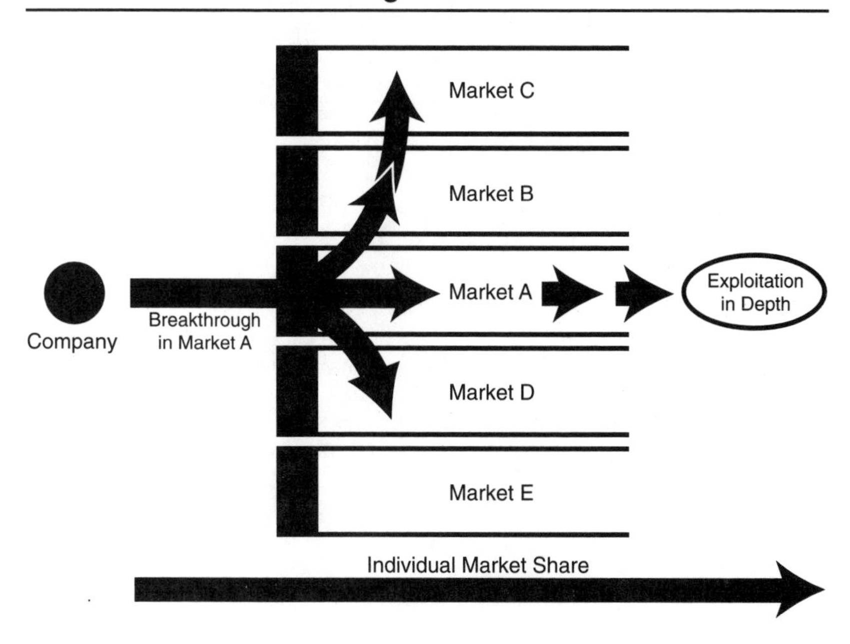

In summary, the important thing is to be self-conscious about how you are going to exploit your breakthrough. Many companies seem to think that they are strong enough to go broadly *and* deeply. Or perhaps it's that they see opportunities in both directions. But these are fundamentally different strategies. Going deeper concentrates resources on developing strength revealed in a breakthrough. It treats what's been revealed as a Hammer—shoring it up, improving it, making it less vulnerable. Going sideways relies on revealed strength as a Pivot—depending on it, using it as a springboard for other opportunities. I suggest, therefore, that it courts folly to mix Pivots and Hammers in your company's exploitation strategy. Pick one, pick it for a good reason, and stick with it.

Notes

1. Personal conversation with Lt. Col. John Shull, U.S. Army (ret.).
2. David Bank, "Microsoft's Huge Cash Hoard Becomes Big Weapon for Entering New Markets," *Wall Street Journal*, 19 July 1999, A3.

3. Charles Schwab with Catherine Freeman, "Success through Failure," *Hemispheres*, November 1999, 54–56.
4. Thermo Electron and Microsoft 1998 annual reports.
5. Gary Lynn, Joseph Morone, and Albert Paulson, "Marketing and Discontinuous Innovation: The Probe and Learn Process," *California Management Review*, spring 1996, 8–37. These authors use the term *probe* in a much broader sense than mine. In their article, it means any attempt to test market reaction with a product, whether as part of a portfolio of tests, as was the case with Searle, or as a single test to be refined by market reaction.
6. Ibid., 17, 25.
7. Ibid., 17, 21.
8. Ibid., 21.
9. Personal conversation with Phil Shephard, preserve manager, The Nature Conservancy, 1999.
10. Robert Leonhard, *The Art of Maneuver* (Novato, CA: Da Capo Press, 1991), 113–15.
11. Ibid., 113–14.
12. Ibid., 115.
13. Ibid., 117–18.
14. In very rural areas, the U.S. Postal Service requires a post office pickup. And now that Roadway Package Express (RPS) has been bought by Federal Express, UPS may be in for some national competition, although to date, RPS seems to be focusing on serving businesses only.

Chapter Eight

Deploying Resources: Tools for Exploitation

> Objective evidence and certitude are doubtless very fine ideals to play with, but where on this moonlit and dream-visited planet are they found?
>
> —WILLIAM JAMES, "THE WILL TO BELIEVE"

Benét's Reader's Encyclopedia defines a *leitmotif* as "a recurrent musical theme that coincides with each appearance of a given character, problem, emotion, or thought."[1] If this book has a leitmotif, it is the insidious perversity of the real world: the tendency of experience to overturn our most reasonable expectations. That's why I have urged that we bring to the forefront of strategic thinking such concepts as hypothesis falsification, the inherent variability of business situations, and courtroom trials.

Against such a backdrop, the appearance of real-world, money-making success—signalling, you may fervently hope, a strategic breakthrough is one of the most exhilarating—and therefore dangerous—moments in a company's history. The reason is that a breakthrough demands the wholesale commitment of resources to its exploitation. But the breakthrough may be profound or superficial; it may go to the root of a market need, or it may be a flash in the pan.

Usually, we can't know where the breakthrough will come, as Searle found out. When Motorola first started exploring the concept of cellular telephones, it commissioned a study to find the top thirty market segments most likely to be interested in the product. Salespeople were the thirty-first most likely group.[2]

There are two important tools for dealing with this additional challenge, the manifestation of the unexpected:

- focusing on asset mobility and
- creating reserves.

Asset Efficiency vs. Asset Mobility

When I cite *asset mobility,* I'm referring to the company's capabilities—its assets—for exploiting unpredicted opportunities thrown up by the multiple-attempt strategic approach I've suggested for an inherently variable world. Asset mobility means that the assets of the company can be used to exploit varying opportunities, but more importantly, the assets can be shifted to exploit new breakthroughs as they become apparent. Not only are these resources flexible enough to be *shifted to* new opportunities, but they can be *withdrawn from* opportunities that have proved less promising. Although this sounds like another glass-is-half-empty proposition, the fact is that when we invest our assets we often think only of the cost of getting into a new opportunity, not the cost of getting out. Further, recall that at the end of the OCE process, a company selectively withdraws resources from its strategic Hammer and redirects them to support the new strategy.

This attitude toward investment and commitment is too uncommon. *Asset efficiency* is more prevalent. It's much easier to justify—to an investment committee or a board—fixed assets that maximize the efficiency with which the task at hand gets done. It's especially hard to argue for flexibility when you, the strategist, can't specify in advance how that flexibility will be used. The future is often uncertain.

In terms of efficiency, an auto assembly plant is about as efficient as it gets. People are hired based on their expertise or the company's ability to mold them into the experts it needs. Plant sites are selected in terms of the lowest cost of utilities, access to low-cost transportation, and the ability to recruit a skilled, low-cost labor force. Equipment is secured on the basis of a cost-benefit trade-off among initial and life-cycle costs, quality, and unit costs. This is as true in the service sector as it is in the manufacturing sector.

Loan generating and loan servicing departments in banks are engineered to evaluate *quality* and *efficiency.* Of course, the criteria for efficiency and quality change as fashions in what constitutes creditworthiness change or as the banks' focus moves from real estate to commercial middle market, or from home equity or mortgage to credit card lending.

The approach called "sticking to your knitting," popularized by *In Search of Excellence* and many other books, encouraged a focus on acquiring assets (including human assets) that were fashioned to support the *core competency* of a company. This approach is excellent, so long as the pace of change is slow and one's expected return on such a specialized investment in specialized assets—material or human—is likely to be measured over a long span of time.

My experience is that there is sometimes a nod to flexibility, but the finance people and the production people usually seek out the lowest unit costs for what they can see. They can put hard numbers on these items. Flexibility sneaks in the back door; often, lead times for plant and equipment are so long that investment decisions must be made before the company knows exactly the shape of what it's going to make. Production and finance may wind up investing in what is seen in retrospect as a suboptimal alternative. So this veil of ignorance sometimes demands that company leaders choose flexible machinery, people, vendor contracts, and so on, because they don't know for sure the exact form and volume of the ultimate product. What is seen today as cost-inefficient and suboptimal is often closer to what is strategically optimal tomorrow.

Consider one of the most modern and efficient auto assembly plants in the world, the plant built by Mazda in Hiroshima, Japan. When it was completed in 1992, it was the most gleaming, automated, robotized, efficient, and modern in the world, and a monument to the superior engineering on which Mazda has always prided itself.

The problem is that it is highly specialized. It can very efficiently make the kind of car Japan and America wanted years ago, but competition and changing consumer tastes have almost forced Mazda out of the market the factory was built to serve. In 1996, this beautiful factory ran at only 35 percent of capacity. Today, it is not quite idle, but worst of all, it cannot make the kinds of cars that are in demand. It is not flexible.

Ford has a substantial stake in Mazda and has a history of joint ventures. The Hiroshima plant albatross was one of the main reasons Ford raised its minority stake from 25 to 33.4 percent, for it cannot afford to have its investment fail. And this is one of the reasons the head of Mazda in Japan is now an American from Ford.[3]

Asset Flexibility

The moral of the story goes beyond the admonition to be careful what kind of plant you build. The moral is that in an inherently variable world, you must seek assets that are flexible without necessarily trading off efficiency. Toyota provides a counterpoint to Mazda: same industry, same culture, different approach.

When Harvard Business School researchers visited Toyota in the mid-1980s, at the height of the (well-deserved) hype about Japanese manufacturing superiority, they were shocked to find few robots. Instead, they found much general-purpose machinery of an older generation, but meticulous attention to manufacturing logistics and process.

For example, to maintain flexibility, Toyota focused on reduced setup time, building special rigs and training its production teams to rapidly adapt its general-purpose machines to make parts for multitudes of models. Thus, Toyota was flexible enough to meet changing dealer needs and customer demand and never locked itself into being able to make only a limited set of models with limited options.[4] Toyota is still the world's lowest-cost and most time-efficient producer, proving that efficiency and flexibility are not necessarily natural enemies.

Toyota certainly shows that flexibility can actually support *long-term efficiency*—that flexibility can be the road to the greatest overall efficiency in performing a service. In contrast, the Mazda plant, which is forced to produce models other than those it was so ingeniously designed to produce, is a paragon of *inefficiency*.

But Toyota appears to have wrung short-term efficiency out of its plants as well. How? By taking aggressive advantage of the most flexible machine of all, the human being. Its production engineers and assembly team workers focused on everything that could be controlled, such as setup times. They created some special-purpose but relatively inexpensive machinery to make up for the inherent inefficiencies of their very adaptable machinery investment. But most importantly, they analyzed the work and reorganized it to make it maximally efficient.

It's the humans who make the machines efficient, not vice versa.

Years ago, I occasionally attended sports car rallies. At one, among the low-slung, fast-cornering sports cars stood a bright yellow Ford Maverick, surely not anyone's idea of a sports car. I asked the rally master whether it was what it seemed to be. Sure enough, it was an ordinary Maverick, recently rented from a major car rental company. It stood still throughout the rally, but for the finale, the best driver was given the Maverick to race through the course. Its skinny tires screamed and the car body rolled terribly on its marshmallow suspension as the driver swung the Maverick through the course. His score? Third place, behind only his own score and that of the runner-up driving true sports cars, but seconds ahead of all the other competitors. Talent trumps technology.

I'm not going to argue that a company should *always* choose the more flexible asset. That depends on the inherent variability of its industry. And efficiency, of course, should have its say. For example, when you are choosing a monster truck with wheels the size of small houses to load coal in the great open-pit coal mines, your alternatives are limited. The commodity nature of the coal business, its long-term stability, and the brute facts of mining constrain flexibility.

Flexibility and Public-Sector Policy Choices

I confess to special worries about asset flexibility when it comes to tax incentive policies in the public sector. States and counties are falling all over themselves to offer tax breaks to large manufacturers. In Alabama, tax breaks worth from $500 to $2,500 per job for a period of five years lured Mercedes Benz to locate a luxury sport utility plant there.[5] What happens when demand for its vehicles shifts due to a change in fashion or a boost in the price of gasoline?

The same question arises with the Faustian bargains that states and counties make with high-tech industries to lure plants and jobs. Integrated circuit (chip) making is a high-fixed-cost, highly cyclical industry. What happens to unemployment rolls when the inevitable downturn hits the chip market? No contractual guarantee is going to keep those plants open or those folks employed.

Flexible Assets: Not Just Property, Plant, and Equipment

The consideration of assets is certainly not limited to physical assets such as plants and equipment. Earlier, I said that human beings are the most flexible machine of all. But it's also true that people can be the most inflexible assets you can invest in.

Today, there is often an emphasis on hiring and training young people. It's true that early training and education often create extremely high capability, and the relative inexperience and energy of youth create a simulacrum of flexibility. If you hire them young, you can train them your way for your company's purposes—at least the purposes you see today. Yet it's worth asking whether this policy creates tomorrow's dinosaurs. Without basic broad, flexible education, all this energy is directed toward digging tomorrow's ruts today. Many of the employees purged in the early retirement orgies of the last fifteen years were young and flexible once. My bet is that the current crop will eventually come to be perceived as just as inflexible as the last crop, but they have not yet reached the age where their inflexibility matters, or shows.

Hiring from the outside—when the *outside* is defined as *within the industry*—is a garden path to inflexibility. Many companies seem to have a habit of simply exchanging executives with one another. Until the late 1980s, for example, the electronic distribution industry had a habit of trading salespeople and executives among companies. Whereas *fresh blood* would have conferred fresh thinking and competitive advantage, this rotation merely heightened competitive intensity. The imported executives knew instinctively how their competitors—their erstwhile employers—would react to any pricing or operational move.

This turned out to be a great opportunity for visionary thinkers such as Steve Kaufmann, chairman of Arrow Electronics. In the mid-1980s, Kaufmann was almost alone in foreseeing the explosive growth in the international flow of electronics parts and realizing that automated distribution (requiring courageous investment) could create competitive advantage. Unfettered by a background in the industry, and supported by a board with vision and a world-class ability to raise money, Kaufmann transformed the cutthroat industry and drove Arrow to the top of it.

When companies in an industry trade executives, it's an eerie reminder of the American Civil War. With very few exceptions, all the key generals were educated at West Point, many studying under the same professors with the same perspective on the Napoleonic Wars.[6] And the major antagonists often knew one another personally. They had seen each other's reactions in battle, and they had served together as small unit commanders in the Mexican War.[7] It is worth pondering whether each side's ability to predict the responses of the other helped prolong the conflict to near exhaustion on both sides.

Creating and Using Reserves

In Chinese, the ideogram for *crisis* is a combination of the characters for *danger* and *opportunity*. There could not be a more apposite description for the situation a company faces when it experiences a strategic breakthrough.

Breakthrough is a moment of opportunity, of course, as the new path beckons. But it is also a moment of danger, for competitors can usually observe the success. A large competitor adopting a *fast-follower* strategic approach may be using *you* for one of its portfolio pinpricks, standing by, ready to pounce. Matsushita's success in consumer electronics following U.S. companies, Sony, and others; Nokia's success following up on Motorola's pioneering efforts in cellular phone technology; and Charles Schwab's gigantic and successful catch-up in online brokerage services all come to mind.

There is danger internally, too, as the new path transmits shock waves through your organization. Those on the front lines of the breakthrough may be euphoric. They are probably overwhelmed with work. They are receiving attention and press. They may be starved for resources. Meanwhile, their one-time compatriots look on in envy, disdain, or fear, learning what it means to be relegated to a Pivot role—or to no role at all.

The breakthrough is thus a time of crisis. It is time to call on your reserves. Next to mobile assets, including flexible people, the second arrow in exploitation's quiver is the concept of a highly capable, highly mobile, highly flexible *reserve*. The concept is best understood through illustration on a geopolitical scale.

The year was 1979. The Carter administration got the shock of its brief life. Against all predictions and hopes, the shah of Iran had been deposed, and Islamic fundamentalist revolutionaries—shouting virulently anti-American slogans—had taken power. On 4 November, students overran the U.S. embassy in Teheran, and the embassy staff was taken hostage.

Late in the year, the USSR launched an invasion of Afghanistan. It was the USSR's first use of large-scale force outside its own borders since the forlorn Hungarian revolution of 1956, not counting the almost bloodless reoccupation of Czechoslovakia in 1968. It was a year of unpleasant surprises for the Departments of State and Defense, not to mention the American people as a whole.

The United States was far less dependent on Middle Eastern oil in 1979 than it is now. But even so, it worried that the USSR might invade Iran to secure its oil reserves for itself and deny them to NATO countries in the event of war in Western Europe. Alternatively, it might invade Iran to protect its southern borders from a stripe of fundamentalism that was only a little less hostile to the USSR's official atheism than it was to American materialism and support of Israel. Last, but far from least, it was conceivable that the USSR might seek to exploit the turmoil in Iran to obtain a Persian Gulf outlet for a blue-water fleet with which to challenge the U.S. Navy.

Recovering from these shocking events and unsettling possibilities, President Jimmy Carter responded with powerful rhetoric. In his 1980 State of the Union Address, he enunciated the Carter doctrine:

> an attempt by an outside force to gain control of the Persian Gulf Region will be regarded as an assault on the vital interests of the United States of America, and such an assault will be repelled by any means necessary, including military force.[8]

What a commitment! But how to back it up? The Carter administration's formula was to scratch together from existing American military forces a new organization, the Rapid Deployment Force (RDF), initially composed of about 230,000 personnel.[9] (For the sake of comparison, the Persian Gulf War eventually involved about half a million Americans, many in logistical roles, and at the height of its involvement in Vietnam, the United States deployed about the same number in Southeast Asia.)

The Rapid Deployment Force—at least as planned under the Carter and Reagan administrations—exhibited the key characteristics of a strategic reserve:

1. It was *held back,* deliberately not committed to any particular place—even at the risk of weakening an effort that might seem crucial at the moment. Thus, during the 1980s, when a Soviet invasion of Western Europe was considered the most important threat to defend against, the RDF was stationed in the continental United States and headquartered in Florida.
2. It was *highly mobile,* so that it could quickly move anywhere in the Middle East (or anywhere else in the world) and quickly move out of it as well, using U.S. Navy sea-lift capabilities.
3. It was in a *high state of readiness,* so it could be effective immediately. Soldiers didn't need to be called up, organized, and trained.
4. It reported to the *highest strategic decision-making levels* and became the U.S. Central Command. Thus it was (for a while) the President's *own* force.
5. It *wasn't huge.* It was small enough to be commanded easily yet big enough to make a difference.

This is a catalog of the most useful elements for exploiting a strategic breakthrough in a corporate setting, with the addition of one more feature:

6. It should be staffed with *highly capable and flexible people* who have the *unquestioned trust of the CEO.*

The reason such a reserve force is particularly congruent to the OCE worldview is that it respects the uncertainty of the strategic process in moderately and severely variable environments. When you don't know exactly how a strategic opportunity is going to shake out or where success will come, and success comes shrouded in ambiguity, you need to apply competent resources quickly and reliably wherever opportunity beckons.

The Mission and Tasks of the Corporate RDF

The mission of such a corporate RDF is to add resources, talent, and energy where it counts most in a breakthrough. In the case of Searle's breakthrough with aspartame in the sugar-substitute market, it would be to multiply the effectiveness of Searle's sales calls on food processors that might use aspartame—so it might include all of Searle's super-salespeople and more sales-force *feet on the street.* Recalling that Searle's success was almost entirely unexpected—occurring among dissatisfied sugar users—the RDF would analyze aspartame's appeal (Are consumers worried about their weight? Their teeth? Diabetes?

Themselves? Their kids?), shape an on-target advertising message, and address any technical and regulatory problems remaining.

In our hypothetical Kmart situation, the RDF would dig into the details of the success of the initial prototype store. One important task would be to try to strip away the ambiguity surrounding the success, sorting out what factors are repeatable in other markets, exactly what the pricing and merchandising selection should be, and what the store rollout and construction plan should be.

RDFs in the Real World

Perhaps the greatest support for this concept comes from the fact that many large organizations already operate *reserves*. They never call them that, but senior executives often rely on one or two teams headed by trusted subordinates to take on the really important (or onerous) tasks: fixing a business unit, opening up the Russian market, handling a public relations disaster. They turn to the same people over and over again.

These teams generally have many of the six characteristics of an RDF. They report directly to the CEO—in fact, like the RDF, they are formed at the CEO's direction. The talent pool they compose is unmatched within the company. They are mobile—both geographically and in terms of expertise. The CEO entrusts them with everything from overhauling a sales force through reducing floor costs to running a joint venture. The teams often have wide discretion, can act rapidly, are authorized to spend lots of money, and are expected to exploit, revitalize, or stabilize business *situations* rapidly—or eliminate them.

The most visible examples of such corporate RDFs are found in turnaround situations. One prominent example is Sandy Sigoloff's famous turnaround at Wickes, the huge furniture outlet and home-improvement center chain. Sigoloff relied on a small team headed by Bernie Kritzer (and others) with wide authority to revitalize retail stores, change their management, or close them.[10] Kritzer and his team spent months living at 20,000 feet in a private jet, flying from store to store at night, overhauling stores during the day.

Another example is Lockheed's famous Skunk Works, a small, high-intensity, autocratically run team charged with creating breakthrough aircraft in very short time frames. It created the famous U-2 in a matter of months, when President Eisenhower, blinded by Russian secrecy and anxious over reports of a huge buildup in Soviet nuclear missile and long-range bomber capabilities, needed new *eyes* over Russia. It created the equally famous RS-71 Mach 3, a reconnaissance aircraft and the first stealth-technology fighter. In each case, Skunk Works operated as a tight, separate team on a focused problem, usually with a sufficient budget. In each case, it created a successful, profitable program for Lockheed in a new area, giving it a strategic advantage over such competitors as Northrop and McDonnell Douglas.

The CEO of a hospital chain I worked with had his own brand of RDF. It consisted of a senior operations executive, a younger staff executive, and a few support staff. Its budget for all practical purposes was unlimited. The hallmark of this group was the energy and analytical perspicacity of the staffer, which complemented the judgment and experience of the operations executive. That, plus the trust of the CEO, empowered the group to exercise stellar and rapid decision making and to commit millions of dollars where and how this small team thought best. The group was also flexible enough to apply the brakes efficiently on its intense work on an operations-oriented problem, turn that problem over to lower-level personnel, and, on a few days' notice, reorient its talent and energy toward a crucial information-technology project. After that problem was corralled, the team was able to turn its full attention to a financial engineering project involving a spin-off of 40 percent of the company's assets. Yet it was always the same core team.

RDF, Hammer, and Pivot

You may gather from my discussion of Carter's strategic problem and my examples that such a rapid deployment force, or strategic reserve, comes in handy not just to exploit breakthroughs but also to deal with crisis, when the company is playing defense. That is only natural, because it is during a crisis that we most often improvise.

But beyond improvisation, the idea of a reserve is particularly apposite when a company is choosing where to play defense and where to play offense. If a competitor takes a jab at a crucial part of the Pivot, an RDF that's already in place can move to shore up defenses. So suppose, for example, that Kmart were pivoting on certain southern markets where it continued to be profitable, while it launched its new *bright-store* strategy. If Wal-Mart suddenly announced new stores in those southern markets, Kmart's hypothetical RDF might temporarily take over the running of the southern region, to blunt the Wal-Mart offensive.

Reserves in War

Since Napoleon, a key military concept has been the idea of holding a force in reserve. The United States and the West as a whole have utilized the concept extensively, even applying it at small unit levels such as platoons, companies, and squads of ten or eleven soldiers. On the other side of the coin, the absence of a reserve always creates a crisis, as it did for Eisenhower in World War II. Few people realize that the moment came during the liberation of France and the invasion of Germany when we ran out of soldiers with which to pursue the not yet beaten Germans.[11]

Even at its worst moments of crisis on the eastern and western fronts, the German army kept some force in reserve, to stabilize the worst breakthroughs. It regarded this as so important that commanders would deliberately thin their front lines in order to keep reserves of their best troops. They knew that with the numerical superiority of their enemies, especially on the Russian front, a local breakthrough was inevitable. Faced with that sobering fact, commanders used the front-line troops to buy time to identify the main axis of the Russian attack and slow the Russian advance, and then used the reserve formations to plug the worst of the holes.[12]

Creating Financial Reserves and Driving Wall Street Crazy

Keeping an RDF of *talent* is one part of the reserve equation. The other part is *financial.* Financial reserves are not profits creatively squirreled away by CFOs to dress up balance sheets and smooth earnings—although, as you'll see, that kind of exercise might play a role in the building of financial reserves. And financial reserves aren't funds saved for a rainy day. In the mid to late 1990s, American automakers stashed away tens of billions of dollars in cash in anticipation of a sales slowdown and global overcapacity. Their goal is to be able to weather the storm and keep product development on schedule, even though current operations may be hemorrhaging cash. That's financial prudence, not financial reserves.

Financial reserves are more like this:

In the early 1990s, Harvard Business School (HBS) was considered a laggard in research and teaching methodology among those in the top tier of business schools. In its teaching, it focused on case method almost entirely—an approach that had served it well since the early 1900s. But the school had a mishmash of information systems connecting departments with faculty members involved in research and development of new case material, and these researchers and departments with students.

Through the early 1990s, Dean McArthur, head of HBS at that time, resisted attempts, requests, and even power plays aimed at revamping and incrementally improving these information systems. HBS got a lot of flak from alumni, faculty, and peer schools. Stanford, the darling of Silicon Valley, moved up in the rankings of business schools, partly through innovative teaching programs and outstanding faculty, and partly through the fact that its students, alumni, and faculty spawned many a local high-technology start-up.

All along, McArthur was hoarding and building financial reserves. Then, in 1995, despite the grumbling of many faculty, HBS swept away in a single stroke its information and case study management system. It wired dorms and buildings with a powerful intranet system and connected faculty and students, alumni and case publishing in a single seamless system. Using an installed network of high-speed modems, students could now access cases from their dorms or apartments, see downloaded video interviews of the executives described in cases, collaborate with other students in virtual study groups, analyze cases on the intranet, and answer professors' questions by electronic essay.

The point isn't the razzle-dazzle of the technology. The point is that with true strategic insight about what was required to modernize HBS and make it more competitive, McArthur resisted the natural human tendency to temporize with the day-to-day demands and compromise with his colleagues, thus frittering away resources. Instead, he formed a reserve by starving current information budgets. He put the school *on defense* in this sector until he could create a decisive, critical mass of resources. He absorbed the criticism and grumbling during the long time this took. Then he committed decisively, swiftly, and completely the entire reserve he had created—not piecemeal, tentatively, or forever awaiting the arrival of the perfect time and technology—to create a breakthrough for HBS faculty and students and for the school's competitive position.

Another example occurred in late 1998, when telecommunications company GTE made a surprise bid for MCI. Wall Street was shocked by GTE chairman Charles Lee's announcement that his company had entered the bidding war against British Telecom and WorldComm to acquire MCI, the second largest U.S. telecommunications provider. GTE's bid was all cash and was valued close to WorldComm's stock offer.

The Street was shocked because Lee had developed a reputation for moving very cautiously. The company had built a solid balance sheet, as well as strong long-distance networks and Internet capabilities. But it was not the major player the Street had been thinking of in terms of an MCI deal. Yet the proposed deal made strategic sense, since GTE's twenty-six local phone monopolies, its growing international business, and its nascent Internet business would be complemented by MCI's own fiber-optic backbone and its long-distance network.

In the end, WorldComm won the bidding war for MCI. But what

is interesting to us is that Lee, seeing the strategic fit, was willing to commit his reserves to make the acquisition. With the acquisition, GTE would have gone from being very conservative financially to being highly leveraged and just able to make the interest payments in the first year of operations. Lee was able to make this kind of commitment only by husbanding GTE's financial resources for many years, putting GTE in a position to strike full force when the opportunity arose.

Kicking the Yearly Budget Habit

Because of our planning habits, our stomachs tend to churn at the very thought of a strategy such as Lee's. One reason is that we are wedded to the idea of *yearly* sales and capital budgets. If you think in terms of a yearly cycle, it follows almost necessarily that you think in terms of what you're going to budget this year for sales, what capital expenditures you're going to allow this year for plant modernization, and what you're going to allocate this year to R&D. This habit inevitably leads to a frittering away of cash flow on finite, easily defined projects. Where would GTE have been if every year it had disbursed its *excess cash* in small dividends or made modest little acquisitions? It would have been in no position to make any strong bid for MCI.

Here's another example: A large distribution company I've worked with has spent the last two years buying much smaller companies, using up its debt capacity, diluting its stock, reducing its cash, and fragmenting its management time and talent on relatively trivial acquisitions. Recently, the company had an important opportunity to forge an alliance with a large foreign distributor and create a seamless, continentwide distribution solution for its customers. Guess what? There were no capital reserves left, no management capacity left. All had been frittered away on nonstrategic areas where the company could just as well have played defense. The result was that the financial aspects of the proposed strategic opportunity were laden with difficulty, perhaps requiring another nail-biting trip to the capital markets or even more high-interest debt.

It's difficult to husband financial resources deliberately for a long time and then expend them in a controlled, directed fashion over a short period to seize a strategic opportunity. That's especially true when you can't specify to your board of directors exactly what that opportunity might be. Shareholders and Wall Street analysts—not to mention corporate raiders eager to get their hands on your cash—are liable to complain vocally and publicly if you don't return *excess cash* to the shareholders.

A second nasty habit stems from the conventional wisdom that every company has an optimal capital structure—a harmony of debt and equity that balances cash flow, taxes, dividends, and capital investments. Conventional financial strategy assumes that there is one optimal capital structure—within a range, of course. This depends on the cost of equity, the cost of debt, the prospects for the industry, and other such factors—factors assumed to be fairly constant.

But in a dynamic world, this is nonsense. There is no reason why a company's financial structure—its balance sheet—should be static or even *tend* to be static. It all depends on what the strategy of the company is and what resources are going to be required to successfully accomplish what needs to be accomplished when action is required. So, for example, GTE could properly go for broke, turning its conservative (low-debt) balance sheet upside down for the strategic opportunity the acquisition of MCI offered.

The value Wall Street assigns to your equity is just one more constraint, like a constraint on how much you can charge for your product in the marketplace. But it has only a tangential relationship to the need to concentrate overwhelming resources on a promising strategic breakthrough.

The pale imitation of this idea is something we see regularly. Companies sell off *nonstrategic* assets or businesses to raise cash for a major thrust. This is an emergency version of what I am advocating, which is the deliberate building of potential energy to be released in the service of a testable and tested hypothesis. Lockheed-Martin is in the throes of just such a disposal of noncore assets as its stock price sags alarmingly as of this writing.[13]

Adopt the idea of financial reserves and drive Wall Street bonkers.

Expending Financial Reserves

It's useful to think of financial reserves as the *war chest* of the CEO and the RDF. The job of the RDF is to exploit strategic breakthroughs by pouring resources where success is transpiring with a strong strategic hypothesis as its underpinning. The RDF, with its financial reserve checkbook, helps us avoid our usual habit of adding resources in exactly the wrong way.

For example, the typical advertising budget for a product launch is set at a fixed amount. If the launch goes well and sales rise as expected, the advertising budget is spent as budgeted. If it goes ill, the budget is cut and the launch stalls out. This is almost the opposite of what should happen.

I have no quarrel with the idea of a budget as such. But if sales are succeeding (and we have a genuine hypothesis that is testing positively and getting progressively more precise), it generally makes sense to raise the advertising budget—in effect, committing the marketing reserve to exploit the opportunity. That money will have to come from somewhere—from areas where breakthroughs are not happening and we have to play defense for the time being.

If done correctly, an analysis of the initial sales success tells us why a product and a marketing approach are succeeding, thus sharpening the hypothesis. In almost every case, the initial launch hypothesis turns out to be incorrect in some details and correct for the wrong reasons in others. Then the RDF arrives with its modestly plump checkbook and adds to the already successful advertising campaign. The new advertising is even more effective because the RDF has refined the campaign on the basis of new testing and the sharper hypothesis it produced.

But probably the bulk of new product launches (as well as other strategic initiatives), at least by big companies, are neither wild successes nor bombs. As a result, ad budgets are spent as scheduled, sometimes to the penny. This would astound an ordinary practicing scientist. Rarely do the results in a true scientific experiment—especially one involving living things, which is, after all, what a marketplace is made up of—match quantified predictions with any great precision. Rather, hypotheses are refined, and predictions get closer to results.

How likely is it, then, that a strategist or marketer would hit the nail on the head a priori with a preset $500,000 or $10 million (or another nice round number) product launch figure, with precisely the message and images conveyed by the advertising geniuses? The odds don't invite betting.

All this suggests that it makes far more sense to think of product launch budgets (and budgets for other strategic endeavors) in terms of *front-line budget* and *reserve*. Commit your reserves as and when the hypotheses generated by front-line results yield analytic insight. Use the initial launch sequence to find out what market segments are most receptive to what messages and what product features. Reshape and reformulate your reserves based on the feedback of the first wave. Above all, do not expend all resources in the first wave—on an a priori formulation of what you think the market wants and how competition will respond.

Creating an Extended Balance Sheet

Creating what I call a company's *extended balance sheet* can help you develop a strategic financial reserve. I call it *extended* because it typically includes assets that the Financial Accounting Standards Board does not require to be reported.

The first source is any cash in excess of what is needed to run the company on a day-to-day basis with *all functions playing defense.* This includes any cash in the forthcoming *strategic term* that the company might throw off and that could be added to the cash balance. This is the minimal company. Remember, any padding in this or that function is an implicit acknowledgment that the function is playing offense. Strategy comes first, so quiz that manager: "How much less than that can you get by on? You're playing defense in today's budget go-round."

The second source is unused capacity for floating equity in the capital markets. It might mean a higher stock price than the industry average, allowing acquisitions of other companies with stock. In the bidding for MCI, upstart WorldCom—with revenues of only $7.4 billion but a stock price with a price-earnings ratio of 75—could announce an acquisition bid for the much larger MCI, with its sales of $19.7 billion and a price-earnings ratio of 22.[14]

Other places to find unused financial capacity include the following:

- Unused or salable property, plant, and equipment (PPE). This includes not only PPE that can be sold for cash, but also unused space and machinery not needed for functions playing defense.
- Budget amounts for service and overhead functions above the *playing-defense* threshold.
- Money from talent not needed for playing strict defense. Not only the budgeted money goes into the strategic reserve kitty, but also an inventory of the talent's capabilities, accomplishments, and ideas. (And it wouldn't be a bad idea if the strategic reserve contained an inventory of the *excess time of the talent* over that required for playing defense. This can be measured in dollars, although a thorough inventory would keep a record in much more than dollars and cents.)

Figure 8-1 illustrates the three kinds of reserves.

Practical Implications

Breakthrough and Exploitation Approaches

Here's a framework for choosing your breakthrough and exploitation approaches:

1. To counteract the natural tendency to attempt a strategic breakthrough with a drive-and-concentrate approach, start by *ruling that out*

Figure 8-1. Exploiting a breakthrough rapidly involves creating and committing reserves.

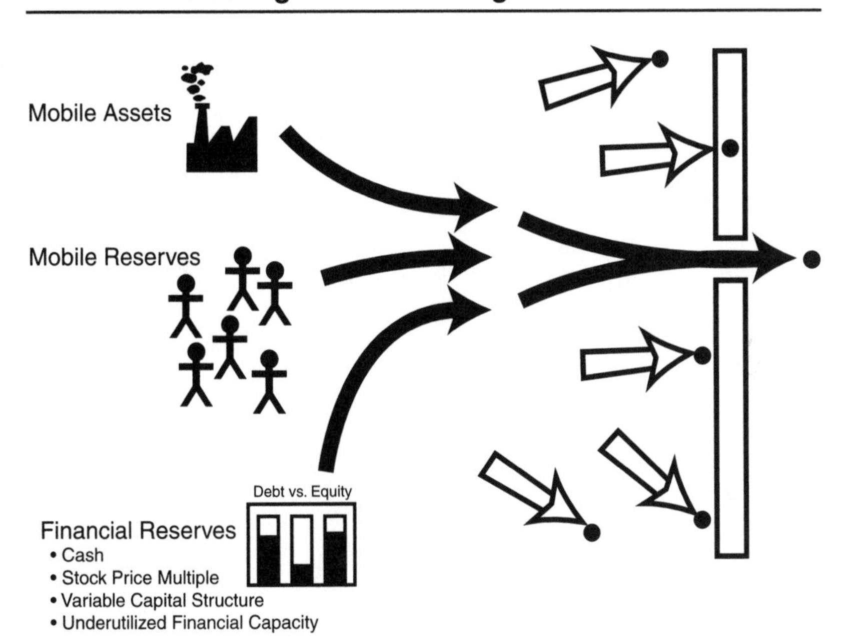

as a possibility. This means stretching your mind and your colleagues' minds in a brainstorming session with the following agenda:

a. Develop as many different ideas as you can to field a portfolio of pinpricks at modest or low cost. How can we substitute simulations, prototypes, proposals, or announcements (for example) to search for the place, time, and form of our strategic breakthrough before we bet the company? Consider using a guided-imagery retreat to provoke some new ideas.

b. Develop as many different ideas as you can for using something besides the company's cash to field the simulations, prototypes, announcements, or whatever. Pretend, in other words, that it is your own money and you're working out of a garage: What can we borrow, rent, or lease? Who else has a stake in the outcome and will invest money, machinery, or time in our test? (General Foods did it with aspartame for Searle.) How can we reward the fielding of a multiplicity of ideas quickly, at low cost, in the atmosphere of a Renaissance festival or county fair?

2. If this doesn't provide a full panoply of pinprick approaches, pull back to the drive-and-concentrate approach *gradually* (no throwing up of the hands, please). Pull back gradually by reducing the num-

ber of pinpricks from numerous to a half dozen. Pull back gradually by using a few alternative sources of money rather than throwing in the towel and trying to shoehorn the project into the yearly budget. Give in, but reluctantly, to that *one best solution*, drive-and-concentrate impulse.

3. When it comes to exploitation, think: With this precious, jewel-like breakthrough in hand, what does it mean to exploit it by going deep? Do we strengthen ourselves with depth, really refining and understanding this success, reinforcing it through steady improvement? Do we make this Hammer into a nearly impregnable Pivot for the future? Or is this opportunity so ephemeral that we must go sideways, applying the breakthrough's advantage broadly but thinly to other markets or nearby technologies or foreign geographies? Start here:

a. Really understanding what this breakthrough means requires an analysis of ____.

b. I expect the opportunity this breakthrough represents to last:

_____ Only a few months (or other short time compared with our investment in it—like a trendy fashion) with a tremendous crash at the end.

_____ Only a few years, but with a graceful fading away as demand wanes or competitors enter the game.

_____ For more than a few years and for a very long time compared with the investment required.

c. The ready application of this breakthrough is:

_____ Fundamentally copying and extending this success (e.g., by applying it to other, similar geographies).

_____ Analogous to what I'm doing already.

_____ Nothing like what I'm doing already, but capable of supporting it competitively (as in Compaq's move into adjacent computer markets).

d. Based on this breakthrough, if I keep on doing what I have been doing, in a few months [years, decades], the Hammer of my strategy will look like ____________ and my Pivot will look like ____________.

Corporate and Financial Reserves

We have identified numerous important practical implications from this study of corporate reserves:

1. First and foremost, consider creating a rapid deployment force: a few highly competent, trustworthy, and well-budgeted team members dedicated to helping front-line staff and able to pour on the coal when a strategic breakthrough needs to be exploited. Don't squander this talent on seemingly crucial efforts that are not panning out but

reward them tremendously for exploiting breakthroughs that show promise.

2. Inventory the flexibility of your human assets by conducting job interviews with your *current staff.* Be ruthlessly objective but dig deep. You will find a few who are more flexible than you thought—the controller (who is a marathoner), the marketer (who is a jazz musician), the public relations man (who had two tours in Vietnam)—but only if you give them the opportunity to show it. It used to amaze me how flexible and broad people are outside of work. Then I realized that they often check their total selves at the door, assuming that the environment inside is command and control. And they wait for orders.[15]

3. Ditch the annual budget and capital investment cycle, at least when it comes to your strategy reviews. Strategically important projects weigh in at their own time. Annual planning is a Wall Street necessity, but not a driver. In some industries (software, for example), eight weeks may be the appropriate strategic horizon. In pharmaceuticals, eight years or even eighteen is more appropriate.

4. Repeat your cash flow and balance sheets to identify total financial reserves over the strategic time line.

5. In your extended balance sheet, identify what assets are flexible and how flexible they are. Match the assets—people, machinery, processes, or plants—against the variability of your environment. Seek out the Mazda Hiroshima plants in your company. They may be human. Check and see how good the match is. Then compare them with the liabilities that will inevitably come due, those that can be restructured, and those that will come due if your current strategy or the one you're implementing doesn't pan out exactly as expected.

6. Don't be fooled into thinking that there is a single optimal capital structure for your company or your industry. There is only an optimal capital structure for the *preserve commitment and postreserve commitment* phases of your work.

7. Create *reserves of time* by getting small, hypothesis-testing projects started early, building slack time into projects, and making decisions earlier rather than later.

8. Be ruthless about cutting losses when hypotheses are not being supported by experience. Exploit success only. Play defense elsewhere.

9. In your annual budget cycle—for you must have one—play defense everywhere except where you are testing or exploiting strategic hypotheses. Don't let each department make up its own budget, because what is optimal on a yearly basis for each department is suboptimal for the enterprise as a whole.

Notes

1. *Benét's Reader's Encyclopedia* (New York: HarperCollins, 1996).
2. Gary Lynn, Joseph Morone, and Albert Paulson, "Marketing and Discontinuous Innovation: The Probe and Learn Process," *California Management Review*, spring 1996, 15.
3. *Business Week*, 7 October 1996, 22.
4. William J. Abernathy, Kim Clark, and Alan Kantrow, *Industrial Renaissance* (New York: Basic Books, 1983), and personal conversations with Abernathy and Kantrow.
5. Albert W. Niemi, "Sweetening Pot for New Industry," *Georgia Trend*, August 1999, 12.
6. The exceptions were often so-called political generals (such as "Butcher" Butler), who were commissioned in the North by Abraham Lincoln to appease this or that constituency.
7. Herman Hathaway and Archer Jones, *How the North Won: A Military History of the Civil War* (Chicago: University of Illinois Press, 1991).
8. "State of the Union," 23 January 1980, in *Public Papers of the Presidents: Jimmy Carter, 1980–1981*, quoted in Joshua M. Epstein, *Strategy and Force Planning—The Case of the Persian Gulf* (Washington, DC: Brookings Institution, 1987), 2.
9. Ibid., 2–3.
10. Personal conversations with Bernie Kritzer and Alan Kritzer.
11. David Eisenhower, *Eisenhower at War, 1943–1945* (New York: Vintage Books, Random House, 1987), 483–85, 525–26.
12. "German Defense Tactics against Russian Breakthroughs" in *Operations of Encircled Forces* (Washington, DC: Center of Military History, U.S. Army, 1988).
13. Chairman's letter to shareholders, October 1999.
14. WorldComm 1996 annual report; MCI annual report. MCI's price-earnings ratio is for 1996 due to uncharacteristically low earnings in 1997.
15. This just happened again, as I was writing this. The vice president of customer service is just waiting for someone to *give* him an advertising plan so that he can plan the staffing of his call center—instead of going out and getting it from the senior vice president, or declaring what he can and can't do, or hiring an outside call center firm as backup, or making it up.

Chapter Nine

Born Allies and Sworn Enemies: Corporate Strategy Meets Corporate Culture

The good reasons for which we act and by which we account for our actions are not the real reasons.

—GEORGE HERBERT MEAD

Where all think alike, no one thinks very much.

—WALTER LIPPMAN[1]

Strategy's Shaky Throne

To succeed, a company needs a robust strategy *vigorously executed.* I've argued that the best way to create a winning strategy is to create strategic hypotheses, temper them in the furnace of falsification, then aggressively feed all available energy, talent, and money into the crucial elements of the strategy—whether focused products, key markets, or crucial functions—while starving the others. Vigorous execution is demanded for all four phases of successful strategy discussed so far: discovery, breakthrough, exploitation, consolidation.

Strategy is the cerebral side of success. The other side of success is the hands-on *implementation* of strategy. As the catchphrase *hands-on* implies, effective implementation of strategy puts a premium on a responsive organization and its culture. But although strategy requires this responsiveness, it rarely gets to *choose* the culture it must rely on. If we look back at the keys to strategic success developed so far, we realize that a successful strategy in the new millennium makes two stringent demands on a company's culture:

1. The culture must have sufficient confidence to be both a bold, fertile creator *and* a stern tester of hypotheses.
2. The culture must be strong while remaining flexible—perfectly prepared to play uncompromising offense in crucial, targeted areas and unyielding, Spartan defense in others, simultaneously.

In short, if a company is to win, its culture must produce organizational behavior that meets the strategic demands of today's marketplace or, perhaps more uncomfortably, tomorrow's. When a culture engenders employee behaviors the marketplace doesn't esteem, it destroys value and, eventually, the company.

Does this happen? You bet. We'll get to some examples later. In contrast, a culture that produces behaviors in harmony with both the marketplace and a tested strategy has positioned itself as a winner. All it needs is the resources and the luck.

So, yes, strategy has primacy over culture. But strategy desperately *needs* culture. Culture holds the keys to behavior, resources, and ultimate performance. Leaders who are "people people" know this from instinct or experience. They know that cerebral excellence in strategy is worthless without the energy of the culture behind it. They know that a less than perfect strategy supported by positive cultural commitment has a far greater chance of success than a brilliant strategic stroke with lackadaisical follow-through. The brilliant leader who looks behind her to find halfhearted, half-competent staff shuffling where she points is the most proverbial of strategic failures.

More common, however, is the competent leader so curbed by culture that an on-target strategy can't get out of first gear: Smith at GM, Ford at Ford, and Volkswagen prior to Piech;[2] AT&T before Armstrong and maybe still; IBM pre-Gerstner and maybe still; the Democratic and Republican Parties, Japan's Liberal Democrats, and Britain's Tories.

So although culture can be the strategist's most powerful and necessary ally, more often it is strategy's most intractable foe. In this chapter, we'll go to the root of the power of culture and discover why culture—without careful husbandry—nearly always sabotages strategy. In the next chapter, I lay out the conditions for creating a hypothesis-testing, breakthrough-exploiting corporate culture and breathe new life into a 2,500-year-old concept that still rocks the world.

Defining Culture

What is culture? Sociologists, anthropologists, and biologists offer diverse, complex answers. Fortunately, for our purposes, we need only answer the question operationally; we need to establish enough about culture's roots to understand its tangible, visible impact on strategy. In the end, what the strategist cares about is what the *tangible products* of a corporate culture will be.

To understand culture operationally, it's useful to contrast good, better, and best corporate cultures, from a strategist's point of view. A company's culture machine creates and sustains a set of employee behaviors. Good companies create sets of behaviors that match customer requirements and meet the company's strategic needs. Better companies create employee behaviors that anticipate evolving customer demands and adapt to shifting competitive conditions. So far, this is nothing new. The best companies create supercharged behavior that goes beyond anticipation to the active development of opportunity—behavior geared to *creating opportunity*.

But a fantastic company has a *machine-tool* culture. Machine tools are used to make tools and parts for tools, such as lathes, drill presses, and milling machines. Their distinctive attribute is their ability to form *many different kinds* of surfaces: holes, internal threading, external threading, cylinders, planes, and points. It all depends on the skill of the operator.

A company with a machine-tool culture can readily create new patterns of behavior in its employees. Sometimes the pattern is prideful and arrogant; sometimes it is detail-oriented; sometimes it is penny-pinching; sometimes it is extravagantly profligate. It may be by turns highly integrated or decentralized; it is a complex of patterns whose constituent elements are removed, replaced, or renovated as strategy and circumstance require. It might play stubborn, money-hoarding defense with a product line that is a little old-fashioned, building cash for a quantum leap. At the same time, it might be spending money in "wasteful" experiments, supporting hungry assistant professors in distant universities, working toward a quantum leap over competitors.

It realizes that no one of these is right for all time. But it realizes that at any given time, many are wrong.

Of course, few companies meet this machine-tool ideal. But now and then, companies exhibit a spectacular cultural flexibility that allows them to choose new required behaviors. Alfred Sloan was able to create a culture in GM's early days that produced a rainbow of behaviors: Sometimes GM was a cost-cutting consolidator, sometimes an experimental marketer of multiple car lines; then it turned into a financial powerhouse and soother of ruffled governments.

More often, however, culture produces behavior that eventually trips, ensnares, and entwines strategy. To understand this—and the remedy—let's look at culture's roots.

The Roots of Culture

How does culture get started? Why does it betray strategy as often as it allies with it? When a founder creates a company, he or she is the organization. If the company is successful from the outset, there can be no fatal mismatch between organization and culture, or there would be no initial success in the first place. Mismatches, however, may be disguised or covered up by the very urgency of the start-up, the success of the product or service, and the slack the first few employees cut the founder in the rush of the kickoff, the first promising sales, and the first perfume of sweet success.

Still, the strains of culture begin with the founder. The company grows up around the founder's vision, personality, and early successes. A company with a strong founder, such as EDS under Ross Perot, Microsoft under Bill Gates, or Ford under Henry Ford, builds a culture in its founder's image. More often than not, the culture endures even when the strategy must change, as happened to Ford in the 1930s. Ford Sr. hated the idea of a six-cylinder engine. While GM was showing the industry that this was what consumers wanted, Ford took an ax to the engine block of his engineers' prototype.

It took the financial and product crisis of the early 1950s to change Ford's culture. In their fifteen years at Ford, Robert McNamara and the Whiz Kids installed a cost accounting–operations–research culture whose vestiges are visible today (and which McNamara took with him to the Department of Defense and Vietnam). Says McNamara in his autobiography:

> To this day [1995], I see quantification as a language to add precision to reasoning about the world. Of course, it cannot deal with issues of morality, beauty, and love, but it is a powerful tool too often neglected when we seek to overcome poverty, fiscal deficits, or the failure of our national health programs.[3]

After McNamara's departure from Ford and the United States' departure from Vietnam, a senior Ford executive said to a junior executive, in retrospect, "McNamara was the kind of man who knew that if you saved a nickel a car on headlights on a million cars you'd make $50,000. But he couldn't tell you why people needed headlights."[4]

Culture can dominate strategy from the very start. It's unusual, perhaps even an abnormality, for a start-up company to be awash in cash. However, I know of one situation like this, reminiscent of the case of the dog that didn't bark. The power that culture can wield over strategy can be illustrated by the strange and ongoing case of Healtheon.

In the mid-1990s, Jim Clark, one of the fabulously successful founders of Sun Microsystems and Netscape, created a health care information company called Healtheon. I visited Healtheon when it was less than a year old. There was something strange about it. Although the company was brand new, with no customers and no revenues—from the elegant waiting room, I could see the potted plants being delivered—the person I met there felt the need to talk with me behind closed office doors. The company already had its share of big-company politics. The culture of a much larger, self-assured, and much less hungry company had taken hold already.

The reason? Clark's astonishing reputation brought with it a ton of venture capital. Healtheon was created in the image of Sun and Netscape *after* they'd become successful—engineering dominated, well populated, and carrying an aura of uniqueness and success from day one. It was definitely not the legendary frugal, embattled start-up with an idea, a few thousand bucks, and a gung-ho salesperson.

But today, Healtheon's strategy has changed several times—first targeting medical records management, then health insurance companies, then physicians and hospital groups, now insurers and employers. After losing $13 million on $5 million in revenue, it acquired ActaMed Corporation (acquisition is another big-company behavior), which already links 60,000 doctors together. That boosted revenue to $75 million, so it could have a shot at an initial public offering (IPO). Unluckily—and even the Jim Clarks of the world need luck—the IPO market cratered, and Healtheon had to pull its IPO off the block.

Meanwhile, Healtheon's engineering talent and the software programs it has written are searching for a marketplace to play in. And, apparently undaunted and unbowed after the first round, Clark raised another $38 million—including $11 million of his own money—to keep the search going. Healtheon's search for a market provides a stick-to-your-knitting parable to end all stick-to-your knitting parables. Finally it merged in the late 1990s with four other medical information companies in a headlong attempt to create critical strategic mass. In *Forbes* and *Business Week* interviews, Clark seems to suggest that roaming the business forest for a place to roost has been Healtheon's core competence.[5]

I call it a luxury. Healtheon's seems to be a culture in search of a strategy.

The Great Schism

If culture diverges from strategy and subverts it, it's because the imperatives of the two are distinct. The imperatives of strategy are external to it—what the marketplace bought yesterday, what it demands today, what it might wish for or need tomorrow. Although strategy may choose to ignore these imperatives for a while or willfully misinterpret them, it must eventually touch the hot coals of fact, especially when cash is leaving the bank.

But culture has its own imperatives. And without early and deliberate intervention, they are congruent with the imperatives of strategy only in the most accidental of cases.

Culture exists for three reasons. First, human beings evolved to recognize and act on *patterns.* The rigors of a million years or more on the savanna made the descendants of the survivors adept at spotting hints of danger and opportunities for food, shelter, and sex. In fact, we are so aggressive in our search for patterns that we seek and infer correlations even when they're unlikely to exist. Experiments show that the smaller the sample of our experience, the quicker we are to infer correlations that may be statistically invalid but may have lots of survival value.[6]

We especially look for patterns of behavior in our fellow human beings, particularly those in positions of dominance. That, too, is survival common sense. Survival requires that we quickly infer patterns from our environment—human or natural—and then quickly create responsive patterns of behavior with survival value. These inferences and their responses spawn culture wherever two or more of us are gathered together.

Second, culture arises from deliberate choices that are deemed *useful,* and they, too, become ingrained. The greatest asset the consulting firm McKinsey has is its culture. Even before new consultants are recruited, McKinsey takes steps to instill in candidates a pattern of behaviors, attitudes, standards of quality, and beliefs. As a result, McKinsey consultants are almost instantly comfortable when they are dropped into new consulting projects anywhere in the world, working with teams of people they've never met and whose native language they may never have heard. There's a high level of confidence about how an analysis will be conducted, how a project will be managed, how fast phone calls will be returned, and what the prerogatives (by seniority) are of each consultant.

It is an astonishing accomplishment that this culture does more than span national or even intercontinental boundaries—it transcends them. German, American, and Japanese McKinseyites have a higher plateau of common understanding and mutual expectations than German, American, or Japanese individuals have with their fellow nationals. And it's practical. This acculturation saves an enormous amount of time in training, getting to know one another, and working out hierarchy and common goals. And it reduces to a tolerable level the amount of interpersonal conflict inherent in an organization of talented, high-energy prima donnas.

McKinsey is far from alone in this. From IBM to the Jesuits to the U.S. Marines, culture has been a deliberate tool dedicated to ingraining automatic human responses. In some environment—perhaps 500,000 years ago on the savanna, or in the 1960s—these tools, these automatic responses, increased the chances for survival and success.

Third, culture is a way of transmitting learning to new generations, be they neophyte consultants, software salespeople, missionaries, or chimpanzees:

> Both chimps and humans exhibit socially learned patterns of activities that last from one generation to the next. . . . For instance Tai chimps eat ants by holding a stick with one hand and dipping it among soldier ants guarding their nest entrance. As ants climb the stick, the apes withdraw the tool; with a hand twist, they sweep off insects with their lips.
>
> Gombe chimps also use one hand to place a stick among the same species of soldier ants. But after withdrawing an ant-laden probe, they sweep it through the closed fingers of the free hand and then shove the mass of insects into their mouths . . . this procedure more efficiently gathers ants than the one observed at Tai. . . . Boesch [a researcher] also reports examples of teaching among Tai chimps. Some mothers leave nuts and stone "hammers" in position near anvils for their infants to use. One mother demonstrated a slowed-down version of nut cracking for her child, and another mother modified her son's positioning of a nut for cracking.[7]

The Web of Belief

So far so good. Cultures create patterns of behavior. But this is where things get sticky. Every organization that self-consciously imposes a culture on its recruits is aiming at one thing: *right action.* Act this way with a client. Act that way with your supervisor. Do this with your expenses, do that in your presentations.

The old-fashioned term for right action is *virtue.* And old Aristotle saw that virtue is a habit. The courageous or completely honest person is not courageous or honest at all once, ex nihilo, but has spent years cultivating those patterns of behavior, usually with the reinforcement and models of other persons. Little by little, these habits become ingrained beliefs about what *right action* is. Eventually, the courageous or the honest person (or, for that matter, his opposite) has a hard time thinking in any other way.

Thus, the acculturation of a person includes a process of habituation to a set of behaviors. These behaviors give rise concurrently to a set of beliefs about correct action. It is this *web of belief,* to use John Dewey's felicitous term, that is so dangerous to strategy.

Why call this phenomenon a *web* of belief? This terminology emphasizes the crucial properties of the beliefs held by a culture. Those beliefs must be:

1. *Interconnected.* A change in one belief or behavior pattern ripples throughout the person's (or department's) behavior and beliefs.

2. *Conservative.* Just as a rip in one part of a spider's web leaves the basic structure and function intact, so, typically, an attempt to change a belief or behavior pattern in people is absorbed by the rest of the webwork.

3. *Structured.* A spider's web relies on definite shapes and joints and is spun according to a definite pattern. To change, exploit, subvert, or reform a web, it is useful and perhaps necessary to know its structure. A web of belief typically has a few basic tenets to which the other beliefs and behaviors connect.

4. *Flexible.* They are adaptable, up to a point. A web can absorb shock by transmitting the force at the point of impact among many points, dissipating the force and often transmuting it. This flexibility supports a web's conservatism. If a new experience or unexpected fact challenges a belief in the web of beliefs, that fact can often be explained away by seemingly minute adjustments in other beliefs, none of which may alter the structure of the web as a whole. A web can even harbor a gaping hole in its structure without serious compromise to its overall strength and resilience.

Consider the following case study in the insidious strength of a web of belief from Stephen Ambrose's book *Citizen Soldiers,* a history of the Second World War. The time is March 1945. The war in Europe is nearly over. The place is Germany, on the front lines. The speaker is Captain F. W. Norris of the Ninetieth Infantry Division. Here is the result of Ambrose's interview with Captain Norris:

> Capt. F. W. Norris of the 90th [Infantry Division] ran into a roadblock. His company took some casualties, then blasted away, wounding many. "The most seriously wounded was a young SS sergeant. . . . He had led the attack. He was

> bleeding copiously and badly needed some plasma." One of Norris's medics started giving him a transfusion. The wounded German, who spoke excellent English, demanded to know if there was any Jewish blood in the plasma. The medic said damned if he knew; in the United States people didn't make such a distinction. The German said if he couldn't have a guarantee that there was no Jewish blood he would refuse treatment.
>
> "I had been listening and had heard enough," Norris remembered. "I turned to this SS guy and in very positive terms I told him I didn't care whether he lived or not, but if he did not take the plasma he would certainly die. He looked at me calmly and said, 'I would rather die than have any Jewish blood in me.'
>
> "So he died."[8]

The SS sergeant's beliefs about racial superiority, his acceptance of the duty to attack even though his cause was lost, and his regard for the value of his own life were all interconnected. His beliefs were conservative: No amount of argument was going to change them. That he might live with Jewish blood and die without it offered no challenge to his belief-set.

His beliefs were structured. They evidently hinged on a few key ideas about race, duty, and patriotism, to which his other beliefs were corollaries. They were flexible. We can't know for sure, but if he had discovered that an excellent soldier in his unit had a distant Jewish heritage, he probably would have had an answer. It's likely that his belief system would have repaired this hole in its web of belief with: "Oh, that just shows how the predominance of Aryan blood can make up for such a taint." No Jewish Defense Force falsification for him.

Perhaps this sergeant was brainwashed by Hitler and Goebbels. Here is another case from the same era:

> In 1936, the Germans laid down the [keel of the monster battleship] *Bismarck* in Hamburg. In compliance with the treaties then in effect, . . . the German Embassy in London informed the British government that the ship would be 792 feet long, 118 feet wide and would mount 15-inch guns. These data were accurate, but the Germans lied about the draft, armor, and horsepower of the ship in order to make the ship's displacement appear to be within limits.
>
> A few British naval analysts believed that the *Bismarck* was a treaty violation, but most apparently did not. One attaché . . . decided that the *Bismarck* had been designed

> with an unusually shallow draft for waters of the Baltic and was therefore mainly intended to fight the Soviet Navy. Another British officer . . . was reluctant to question the reported dimensions of the *Bismarck* precisely because that would have meant Germany had violated the treaty. Indeed, the British Navy's Intelligence Division did not alter its estimate even after the *Bismarck* had sunk and British analysts had been able to examine the ship's log and interrogate its surviving crew. The Admiralty continued to accept its intelligence service's estimate until a year later, when it received incontrovertible information from Soviet intelligence clearly indicating that the ship was, in fact, in violation.[9]

The human need for consistency in our webs of beliefs is so great that we will actually pay to provide ourselves with a consistent set of beliefs, even though the whole package might be fraudulent. This fact is illustrated in a life experience related in Cialdini's *Influence—The Psychology of Persuasion:*

> One night at an introductory lecture given by the transcendental (TM) program, I witnessed a nice illustration of how people will hide inside the walls of consistency to protect themselves from the troublesome consequences of thought. The program claimed it could teach a unique brand of meditation that would allow us to achieve all manner of desirable things, ranging from simple inner peace to the more spectacular abilities to fly and pass through walls at the program's advanced (and more expensive) stages.
>
> I brought along an interested friend, a university professor whose areas of specialization were statistics and symbolic logic. When the leaders called for questions at the completion of the lecture, he raised his hand and gently but surely demolished the presentation. In less than two minutes, he pointed out precisely where and why the lecturers' complex argument was contradictory, illogical, and insupportable. The effect on the discussion leaders was devastating. After a confused silence, each attempted a weak reply only to halt midway and finally admit that my colleague's points were good ones requiring further study. More interesting to me, though, was the effect upon the rest of the audience. At the end of the question period, the two recruiters were faced with a crush of audience members submitting their $75 down payments. Nudging, shrugging, and

> chuckling to one another as they took in the payments, the recruiters betrayed signs of giddy bewilderment. After what appeared to have been an embarrassingly clear collapse of their presentation, the meeting somehow turned into a great success, generating mystifyingly high levels of compliance from the audience. I chalked up the audience response to a failure to understand the logic of my colleague's arguments. As it turned out, however, just the reverse was the case. Outside the lecture room after the meeting, three members of the audience, each of whom had given a down payment immediately after the lecture, approached us. Still thinking that the three must have signed up because they hadn't understood the points made by my logician friend, I began to question them about aspects of his argument. To my surprise, I found that they had understood his comments quite well. The spokesman (for the group of three) put it best. "Well, I wasn't going to put down my money tonight because I'm really quite broke right now; I was going to wait until the next meeting. But when your buddy started talking, I knew I'd better give them my money now, or I'd go home and start thinking about what he said and never sign up."
>
> All at once things began to make sense. These were people with real problems, and they were somewhat desperately searching for a way to solve those problems. They had found a potential solution in TM. Driven by their needs, they very much wanted to believe that TM was their answer.
>
> Now, in the form of my colleague, intruded the voice of reason, showing the theory underlying their newfound solution to be unsound. "Quick, a hiding place from thought! Here, take this money. Whew, saved in the nick of time! The decision has been made, and from now on the consistency tape can be played whenever necessary. TM? Certainly I think it will help me; certainly I expect to continue; certainly I believe in TM. I already put my money down for it, didn't I?"[10]

What is going on here? It's the conservatism of the web of the subject's belief. Paying money is a tangible affirmation of a set of beliefs about the hoped-for miracle. When our author's professor friend points out the logical flaws in the belief, we find that logic is only one strand that supports people's belief and behavior. In this mental con game, the audience's strands of belief include, for example, the belief that there is a (simple, easy, straightforward) solution to their problems; that there is hope in the future; that they didn't waste their time showing up for the presentation; that the earnestness of the presenters was a good proxy for their goodwill; that the reasonableness of some of what the presenters had to say made reasonable the rest of what they had to say.

Of this web, the professor at most succeeded in knocking out only one or two strands of belief—that logic can be trusted and that the path of science is more worthy of belief than that of faith. Perhaps many in the audience believed in the trustworthiness of logic. But they had also witnessed debates (perhaps in law or politics or science itself) in which a logical argument that seemed trustworthy one moment was utterly trashed a few moments later by a skilled debater.

If this is the strength and innate conservatism of the personal beliefs of a small group of strangers, how much stronger and more conservative must be the web of belief in a corporation of dozens, hundreds, or thousands of employees, all acting more or less the same way?

There is a strong impulse in the human makeup to look to what others are doing as an affirmation of what is right. Cialdini uses many interesting and sobering examples to prove that the instinct to look to others is deeply rooted. For example, evangelical preachers are known to seed their audiences with "ringers" who are rehearsed to come forward at specified times to give witness and donations. An Arizona State University research team that infiltrated the Billy Graham organization reported on such advance preparations prior to one of his crusade visits. By the time Graham arrives in town and makes his altar call, an army of 6,000 waits with instructions on when to come forth at varying intervals to create the impression of a spontaneous mass outpouring.[11]

Culture Caricatures Itself

It can get even worse. Inside any organization are built-in *forces* that pressure culture to keep out of touch with a changing reality.

Organizations are built around hierarchy and power. *Power* ultimately means the ability to hire, fire, promote, and destroy. Usually, keeping your job means doing what the boss says. Especially when most promotions are from within, the boss expects subordinates to replicate the behavior they exhibited in their old jobs. After all, as all seasoned executives know, in business, there is nothing worse than a surprise. But surprises are regularly dished out by the world.

Even—or especially—*new blood* brought in from outside is tremendously constrained by existing power relationships. Mr. New

Blood needs to jockey for a position relative to colleagues, establish turf, and establish (as quickly as possible) his competence and credibility. All this means relating to what is already there. It means adopting the very language of the new (to him) environment.

Those habits of language quickly become habits of belief: Mr. New Blood, first adopting them as protective coloration, soon comes to believe in them as strongly as any old-timer. And Mr. New Blood, in his turn, is likely to make the organization even more replete with true believers. After all, anybody who doesn't like the culture leaves or is fired.

Two close friends of mine joined a thirty-year-old policy analysis firm as "experienced hires." Both experienced some trepidation. One came from a political background and was a highly unusual addition to this apolitical firm. The other came from a motley background of academic and sales experiences. The very term *experienced hire* signals that the firm had a tradition of hiring law school graduates who had little law or political *experience*. The tradition was based on the fact that the firm had simply had far better results grooming young, new analyst-lobbyists than trying to remake seasoned executives into the analytical and relationship-building machines that made the firm so successful.

One of my two friends rapidly adopted the protective coloration of the new firm. In retrospect, his key move was finding a mentor who was cut from the firm's traditional cloth. In private, this friend retained a humorous, reflective distance from the firm. But during workdays and business hours, he was a strict believer in the firm's way, truth and light. The other friend, though well respected (having rescued two failing projects), could never find the energy or the desire to adopt the language and customs of the tribe. He was afraid, perhaps, that by buying in he would be selling out.

The first friend is now winding up a full and successful career at the firm. The second was provided a graceful, extended, but definitive exit. He wasn't alone. Gradually, the firm weeded out more and more sources of diversity and difference. In the race for election to partner (this firm is ostensibly a law firm), each candidate is compared with the tinsel ideal of the "perfect partner." This makes the slightest deviations stand out. Those who are most deviant compared with other partner candidates are politely asked to leave. They don't "fit the mold." As the years go by, the firm becomes more like itself than it ever was. That's why I say it becomes a caricature of itself.

Figure 9-1 shows how, over time, a company's culture is likely to lose diversity when—as is the usual case—it's a matter not of survival of the fittest but of survival of those who fit.

Every U.S. Naval Academy graduate who entered the nuclear sub-

Figure 9-1. How company cultures become self-caricatures.

	Company's original cultural diversity	Plurality A's + B's purge D's and C's	A's and B's hire people like themselves with "mistakes"	A's and B's purge "mistakes" and C's	"Mistakes" replaced	Majority A's purge some B's, make converts	Purge victims replaced	The final culture
	A	A	A	A	A	A	A	A
	A	A	A	A	A	A	A	A
	A	A	A	A	A	A	A	A
	B	B	B	B	B	A	A	A
	A	A	A	A	A	A	A	A
	B	B	B	B	B	A	F	A
	A	A	A	A	A	A	A	A
	A	A	A	A	A	A	A	A
	C	C	A	A	A	A	A	A
	A	A	A	A	A	A		A
	C	C	B	B	B	B	A	A
	B	B	B	B	B	A	D	D
	C	C	B	B	B		A	A
	C	C	C		A	A	A	A
	D		A	A	A	A	A	A
	A	A	A	A	A	A	A	A
	A	A	A	A	A	A	A	A
	B	B	B	B	B	B	A	A
	B	B	B	B	B	B	A	A
	A	A	A	A	A	A		
	C	C	C		B	B		
	D		C		C			
	E		F					
Year	1	1	2	3	3	4	4	4

marine service during the days of Admiral Hyman Rickover will tell you that Rickover had in mind a strict image of the ideal submarine officer. It was, all things considered, a fine, admirable, and, above all, dependable one. But a mold it was, and for years you could spot a submariner by the way he walked, talked, and thought. When Rickover passed on, you can guess what kind of people Rickover protégés recruited.

A second force that works to keep culture out of touch with reality is its *stealth.* Culture is so invisible and so subtle and long term in its effects that it can be hard to recognize, especially if you are part of the culture yourself. Culture influences even the most trivial decisions. For example, imagine that an engineering manager is discussing her need for office supplies with the office manager: If the culture of the organization is speed, the multitasking office manager will pick up the phone and call Corporate Express, the office supply delivery company—problem solved (at a higher price).

But our office supply interlocutors will have a very different discussion—and a very different result—if they work in a cautious, multilayered bureaucracy. Chances are, our engineering manager heroine won't get her paper clips in the next two hours. In this company, the office manager reports the situation to the vice president of purchasing, who reports to the senior vice president of administration. A month's delay in choosing an office supply service won't be reflected in his yearly performance review, but a 2 percent increase in the cost of supplies might well be.

The underlying causality is seldom if ever discussed. Who would ever bring up the question of hierarchy in a real, everyday dialogue about file folders and paper clips? Any outsider would be told, "That's just how office supplies are ordered around here."

I once consulted to the CEO and senior staff of a large transportation company. Analysis showed that some fairly straightforward changes were needed to bring some semblance of profitability back to the company. Certain routes needed to be streamlined, certain terminal operations consolidated, certain compensation incentives tweaked. As usual, I mooted these suggestions with a bright, older, savvy manager. He acknowledged the gravity of the situation and the aptness of our team's solution. Intellectually and managerially, he was delighted. But there was something wrong that he couldn't quite put his finger on. For a long moment he looked out into the desert surroundings from his eleventh-story window. Then he turned slowly and said, "Any blade of grass here that sticks up gets cut off." The cultural cash register rang up "no sale."

Here's another case of the invisible power of culture. Once you're hired, you're assigned to an office. There's not much to question there.

You're focusing on your new job, your new boss, your new secretary, the new colleagues you'll need to persuade and rely on, and where the bathroom and lunchroom are. You're just glad to have a desk and an office.

Yet if you stepped back, you would realize that your office can facilitate or impede your success on the job: Your working space influences your culture and performance. At Microsoft's Redmond, Washington, headquarters, the company takes pains to make all offices exactly the same size. You see no cubicles. A whole industry of office systems—Herman Miller and Steelcase, for example—grew up in the 1970s based on theories of influencing culture by influencing office environment. (I happen to think those theories were convenient excuses for saving money on office construction, but in any event, cultural influence there certainly was.

Recently, the Massachusetts Institute of Technology unveiled plans to demolish its famous Building 20.

Building 20 was erected during World War II for scientists engaged in research supporting the war effort. In the harsh Boston winters, it was drafty. In summer, without air-conditioning, it was hot and muggy. The cheap, plain structure had been designed to last for the duration of the war plus six months. It was tired. What MIT didn't anticipate was that as its last tenants were evicted, they would hold a wake for the creaky old building.

Building 20 may be the most famous R&D facility in the United States. U.S. radar inventions were born there. Noam Chomsky revolutionized linguistics there. Groundbreaking computer and brain research took place there. One reason: Since there were no fancy walls to begin with, researchers created the spaces and work areas they needed with whatever plywood, two-by-fours, and Sheetrock they found handy. Water pipes and electrical cables ran exposed along ceilings. Since no one told them they couldn't, professors and graduate students were known to tap into them whenever they needed water or electricity. The place fostered collaboration, camaraderie, exploration, and hard work. Janitors, craftsmen, and handymen downed celebratory drinks with professors and students at the completion of research milestones.[12]

When this creaky monument to freewheeling, high-standard exploration is torn down, how likely is it that the new multimillion-dollar building designed by renowned architects (but with windows that don't open) will become as famous as Building 20 over the next fifty years?

The song says, "You don't know what you've got till it's gone."[13]

Another invisible force of culture might be called the CEO factor. I mean the universal tendencies of underlings to imitate the behavior, attitudes, and even mannerisms of CEOs. Dogs act like their masters. General Douglas MacArthur used to pace ceaselessly during his wartime strategic ruminations while sucking on a corncob pipe. He had a habit of listening carefully to his subordinates' strategic analyses and recommendations in silence. Then he'd speedily dissect the analysis, expose its flaws, and as often as not recommended precisely the opposite strategy. It's well known that many of his subordinates—generals and colonels themselves—became his imitators. With his pipe and his pace, but without his brilliance, these second-rate imitators commanded their troops, and no one can know how many lives they wasted.

My experience has been that many CEOs curse and yell and throw infantile rages. When they do, it isn't long before their subordinates, on the receiving end of this behavior, act likewise toward *their* subordinates. Invisibly, imperceptibly, it works its way down to first-line supervisors. "What's the matter, Jones? What (expletive) do you *mean* you can't keep the customer service people from taking (expletive) sick leave?" It would be fascinating to know how many strikes have had some CEO's toilet training problems as their ultimate cause.

In contrast, when CEOs set standards of civility, grace, and attention to dress, that percolates down too, and very quickly. If confrontation is the norm, as at General Electric, that propagates itself as well. At GE, if you make proposal to a group of executives, you can expect a very rough time. You must learn to repay in kind what you receive. The expectation is overt and deliberate. It stems from the CEO's belief that if a manager won't fight for an idea in a meeting, he or she is unlikely to fight for it in the marketplace. Whether these conference-room slugfests result in truly good ideas is open to debate, but GE's performance is beyond question.

The CEO factor may be hardwired into our brains:

> In monkey colonies, where rigid dominance hierarchies exist, beneficial innovations . . . do not spread quickly through the group unless they are taught first to a dominant animal. When a lower animal is taught the new concept first, the rest of the colony remains mostly oblivious to its value. . . . In one troop, a taste for caramels was developed by introducing this new food into the diet of young peripherals, low on the status ladder. The taste for caramels inched slowly up the ranks: A year and a half later, only 51 per cent of the colony had acquired it, and still none of the leaders. Contrast this with what happened in a second troop where wheat was introduced first to the leader: Wheat eating—to

this point unknown to these monkeys—spread through the whole colony within four hours.[14]

Strategy Bereft

Culture, thus following its own imperatives, routinely sabotages strategy.

I once participated in giving a presentation at Capital Horticultural Services[15] headquarters to show the staff how they might take advantage of the then-new Internet and multimedia personal computing applications. CHS is one of the largest, proudest, and most successful horticultural service companies in the United States. After meeting a wartime need, it made an immensely successful transition to serving the peacetime agribusiness industry and the consuming public.

I was shocked at the response we got from CHS, a response that can be encapsulated by two comments. One was, "We are a data-driven company. We never do anything without reams of market research." The other was, "We are so big and so successful that we don't have to concern ourselves with anything this new. The other people [CHS's competitors] have to follow us."

What I learned from these remarks is that size and success had entangled CHS's senior management in a web of belief impervious to outside perturbations. No logic (in this case, the folly of ignoring the power of the Internet and multimedia agricultural information) could possibly lead them to question their market leadership, competitiveness, or addiction to backward-gazing market research.

This was their attitude despite the volcanic changes in the agribusiness world at large; workers' strikes; calls by CHS's own executives for the removal of its northern California executive officer; the maimed public image of agribusiness in general and, in many places, CHS in particular; and the fact that for the first time in its history, CHS had lost money in 1997—and not a trivial amount. It lost over $250 million on revenues of nearly $15 billion. In fiscal 1998, it lost over $100 million in the third quarter, optimistically saying that it expected a loss for the full year comparable to the quarter billion it had lost the previous year. Then it reportedly rewarded its CEO with a raise from about $1 million to about $2 million a year.

But beyond that, CHS's logical disconnect was that, for most people, the Internet and multimedia were so new that "reams of market research" could not tell them what consumers wanted. You might as well have surveyed Pony Express riders about their experience with the telegraph.

A start-up I recently worked with observed the power of culture to disembowel strategy with subtlety. To rapidly create a sales force for an exciting new product, this venture-funded start-up hired a dozen salespeople, each with ten to twenty years' industry sales experience, from big medical supply companies. The highly experienced regional sales managers brought their big-company ways with them. Hired and trained in these big, leading companies before the automation of sales forces with personal computers, pagers, and cell phones, these regional managers (RMs) resisted tooth and nail the introduction of such "time wasters" into the brand-new 1998 sales force.

So the trap was set. The company's sales and marketing plan depended on recruiting leading distributors that this sales force would call on. On Monday, one RM flew off to meet with a potential distributor. On Tuesday, the company president awoke to find that the company had been committed to a distributor he didn't know and thousands of dollars in ad support that wasn't budgeted. The RM's excuse: He was traveling and had to drive to catch a plane and thus couldn't call headquarters in time. No cell phone, no discussion, no approval. Unhappy CEO, unhappy RM, unhappy vice president of marketing, unhappy distributor. No sales.

The result, of course, was a new policy that distribution agreements and ad support contracts required company headquarters preapproval. The RM's big-company culture, far from being left behind at his farewell party a year before, had bitten the heels of the company's sales and marketing strategy. Result: evisceration of speed, this start-up's crucial asset in a world of much larger competitors.

Must It Take a Crisis?

I don't believe that it absolutely, positively takes a disaster or a crisis to put a company culture back on the strategic rails. But sometimes it certainly *seems* that way.

Look at Columbia/HCA. For years, this hospital behemoth inculcated key values in the minds of its hospital chief executives: double-digit profitability growth, nationwide standardization, anything-it-takes recruitment of doctors, ruthless competition in every market. Headquarters set rapacious annual profit goals for each hospital. Hospital chief executives who didn't make the goals were publicly castigated, not infrequently humiliated, and often fired. No wonder hospital chief executives aggressively charged Medicare, raided other hospitals for doctors, linked doctors to revenue-increasing laboratories, and built facilities designed to drive other hospitals—often local, community-based nonprofits—out of business. This was second nature in Columbia; this was its culture.[16]

Meanwhile, what was happening in the marketplace? Overcapacity—too many hospital beds—hung over many markets. Managed care sequeezed hospital pricing. Medicare started to run out of money, triggering congressional investigations of its fiscal stability (and causing Medicare bureaucrats to examine reimbursement policies carefully). Patients were becoming more knowledgeable, vocal, and critical of hospital care. Politicians at all levels were beginning to listen to constituent complaints about care denied, delayed, or uncompensated.

Yet Columbia's behavior not only didn't change to meet the new environment, it became exaggerated. As it became harder to sustain double-digit profit growth year after year, Columbia's executives accelerated their efforts. Some hospital executives, it is alleged, sought to up-code Medicare reimbursement requests. *Up-coding* means finding the highest plausible reimbursement code for a procedure. In addition, Columbia aggressively pushed capacity expansion or acquisition wherever possible. Hospital executives worked seven-day weeks, cutting costs and courting doctors with high patient admission numbers. Columbia helped its doctors invest in home health care agencies to which, it is said, they could steer discharged patients—health care agencies in which Columbia and the doctors themselves had invested.

Everyone knows what happened. The roof fell in when Medicare officials became aware of the apparent up-coding. The dirty laundry of Columbia's culture spilled onto the front pages of the nation's newspapers and was displayed in color on the evening news. State attorneys general started looking into other charges of fraud and abuse.

Columbia wound up firing its aggressive, deal-making CEO and bringing back Dr. Thomas Frist. Frist shuttered Columbia's expansion program; dumped a large home health care business; put dozens of hospitals, surgery centers, and medical office buildings up for sale (for good prices); and prepared to spin off to long-suffering shareholders hospitals Columbia couldn't sell. Frist changed the internal management incentives. He launched a huge cultural revolution, making community relations rather than national standardization the keystone of hospital successes. He launched internal reviews and investigations and fostered cooperation with the federal investigators. He set up a special ethics panel to review executives' and managers' decisions. He eliminated Columbia's annual employee cash incentive program. He even had Columbia file a $10 million lawsuit against a former senior vice president alleging "a pattern of racketeering."[17]

What's interesting for us is that the inertia of Columbia's original culture had deviated so radically from serving the marketplace that ultimately paid its salaries. The forces of the culture—and the resulting expectations on Wall Street—drove Columbia down a strategic dead-end road to a near-total breakdown of contact with marketplace realities.

Now the question for Dr. Frist is whether these radical cultural and organizational changes—and the strategic changes they imply—can work quickly enough to save the company from acquisition or dissolution. Income from operations in late 1998 was running at about $750 million on $14 billion in sales, about half the 1997 level. But after a dip in same-facility admission when the scandal broke, admissions at the end of 1998 were up slightly.

My friend Robert Neely would say that Frist's actions illustrate what I call *Neely's dictum:* "The first step in change management is to change management."

It will be interesting to see whether it takes a crisis or just a slightly farsighted perception of grim necessity to force a change in the U.S. automakers' web of belief. It's stunning to watch. The Big 3, after decades of resistance, have all decided almost in the same instant to start developing high-efficiency, low-emissions engines.[18]

Since the end of World War II, the Big 3 had been acting within a web of belief that stated that environmental concerns were essentially extrinsic to their core business. Resistance to public need had indeed been the Big 3 pattern. In the late 1960s, there was the fight over lap belts, then the fight over three-point seat belts, then the hard-fought compromise over safety standards and mileage and emissions.

The signing of the Kyoto Global Warming Treaty in 1997 signaled the unraveling of this web. What accelerated it was Toyota's simultaneous announcement of the first production-ready high-mileage, low-emissions hybrid internal combustion–electric car. These events signaled that the world—not just the U.S. Government and the U.S. public—was preparing to intervene directly in the car business. The Big 3's single largest competitor and the world's low-cost producer was stealing a march on them. And low gasoline prices—a key structure in the Big 3's web of belief—wouldn't last forever, because rocketing demand from developing nations, the thirst for new tax revenues, and environmental regulation were likely to drive them up.

In mid-1998, the Big 3 seemed prepared to revamp their web of belief. Neil W. Ressler, Ford's vice president for advanced vehicle technology, professed that the oil crisis of the 1970s had made a lasting impression in Detroit: "The lesson from that is that things that look immutable can change very quickly. . . . You have to be looking ahead and thinking about what happens if that and that. One of them is, what happens if there is an oil crisis?"[19] Harry Pearce, GM vice chairman, chose words aimed at changing employees' beliefs: "We need to do it. We want to do it. And we're going to do it. We're deadly serious about it."[20]

But it's *hard* being green. Less than a year after Ressler's remarks, Ford decided to introduce the Ford Excursion. It is the largest passenger vehicle on the road (10 inches longer than the reigning heavyweight champ, the Chevrolet Suburban), gets 12 miles per gallon (but avoids federal fuel economy requirements), is 6 feet tall, and "will be particularly deadly when it slams into much smaller vehicles in traffic accidents."[21]

The reason, of course, is that the Excursion is expected to be immensely profitable. Even if only 50,000 are sold each year (a minuscule production run by Detroit standards), pretax *profits* will be $1 billion annually. Costs are low because the Excursion is built on an existing frame. Prices are high because Americans, craving sport utility vehicles and lulled by low gas prices, are willing to accept top-dollar price tags for an image they crave.[22]

Does it have to be this way? I don't think so. But changing *the way it is* will take hard work from the top management team (TMT).

Practical Implications: Taking a Cultural Inventory

1. Tomorrow morning, take an operational inventory of your company's beliefs and cultural outputs. You might use a chart like Figure 9-2. I've filled in sample responses based on those from executives of a real company I worked with in 1996.

2. Check the chart. Do a private evaluation with the door closed.

- Where is our strategic center of gravity? (See chapters 1–3.)
- Do the outputs and beliefs really match the strategic needs? How well, on a scale of 1 to 5?
- In our real cultural outputs, where are we playing offense? Are we meeting our goals there?
- Where are we playing defense? Are we meeting our goals there?
- Where we're falling short, what beliefs are the key people operating on? How does what they say match their operational outputs?
- How am I doing? What are my real outputs? What are my real beliefs? What do I need to change?

3. Crucially, ask yourself *why* for each question. And think especially about the following:

- What forces—visible and invisible (power structure, language, routines, CEO factor, beliefs and habits brought in from the outside)—are making things happen this way?

Figure 9-2. Cultural inventory: Smith and Wesson Personal Computer Corporation.

In This Function	*Our Key Strategic Success Factors Are*	*Our Actual Cultural Output Is*	*Because We Believe*
Mfg./Operations	Meeting product introduction date at target cost	We get it out the door when the dealers call	Our real deadline is customer's ship-by date
Engineering	Reliable products fast to market	We nearly match leading competitors	A few really talented folks make anything happen
Design	Elegance in form factor	We get awards	Industrial design awards count
Customer service	Keep customers happy	We don't know, but we think we're pretty good	Our customer service staff is overloaded
Finance	Raise the next round of capital	Our Asian partner kicked in	We must invest and take losses until we build volume
Control	Report costs and margins accurately	Our margins have been very poor	Pricing and margins are up to top management
Human resources	Hire the best people; cost unimportant vs. quality	Turnover isn't bad for this industry	Realistically, new hires see us as highly risky
Sales	Desperately retain our largest customers	Largest customers retained	Without them, we'll never cover our overhead
Logistics	Get suppliers' products on time	We document that we have the lowest prices from shippers	Our budget for shipping costs is tight
Marketing	Price is right; awareness is less of an issue	We've got a great name out there, so we price a hair under the big boys	We really are price takers, not price setters
Information systems	Finish prioritized projects on time	Our backlog is only three or four weeks	We must do the best with the programming talent we have
Top management	Make the right decision; keep Wall Street happy	It takes a strong hand and good gut instincts in this fast-moving place; fortunately, we're up to it	We need to be quick on our feet; by the time we finish analysis, the window of opportunity is gone

- Can these forces be changed, realistically? Rechanneled? Eliminated?
- What new forces and structures are required?

4. Give your colleagues a blank copy of a chart like Figure 9-2. Compare notes. Keep notes. What do these guys really believe? What beliefs are connected to others? I recommend that you read a remarkable resource at this point to get acquainted with an excellent thinking tool called a *mind map,* invented by Tony Buzan[23] (see Figure 9-3).

5. Decide which beliefs—and their holders—need to change.

It goes without saying, I suppose, but *be honest.* The company sketched earlier (see Figure 9-2), a former industry leader, needed this cultural analysis *before* it started losing a quarter billion dollars every three months on sales of less than $2 billion. At this writing, the former CEO is looking for work.

■ ■ ■ ■ ■

I believe that strategy-driven, culturally responsive companies can be made real in today's world. Who pulls on the tiller of culture to steer the corporate ship in the right strategic direction? That has got to be the company's top management team. The next chapter explores

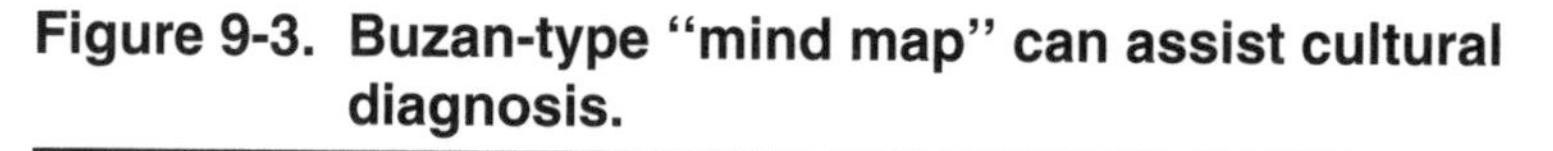

Figure 9-3. Buzan-type "mind map" can assist cultural diagnosis.

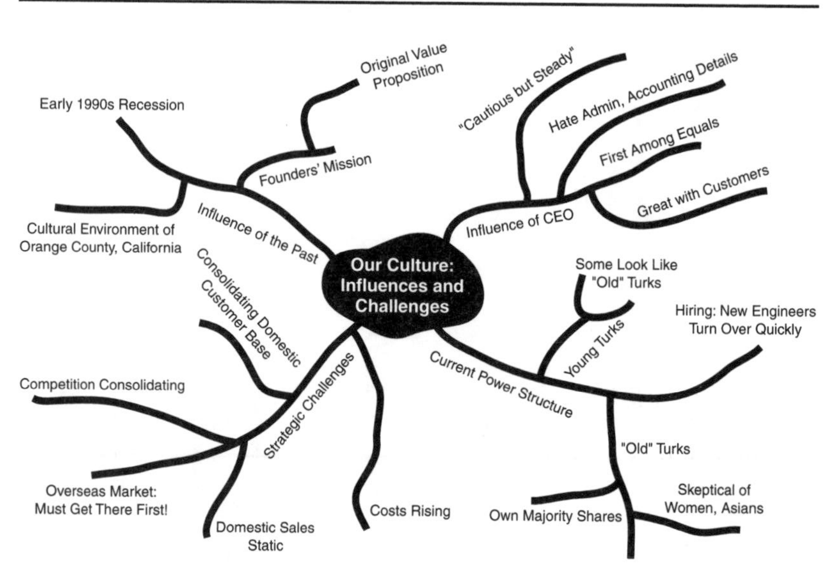

what it takes to build a top management team that's strategically responsive and strategically effective.

Notes

1. Quoted in Robert B. Cialdini, *Influence—The Psychology of Persuasion* (New York: Quill, 1993), 114.
2. A great example of how enshrined, government-certified culture can vitiate strategy is in Paul Klebnikow, "Bringing Back the Beetle," *Forbes,* 4 June 1997. The article describes how Piech turned Volkswagen around, boosting sales by two-thirds, launching the new Beetle, and cutting models to gain efficiency. "But there's an underlying problem that is probably beyond even Piech's talents. Out of VW's 243,256 employees, 57 percent are in Germany. These workers get six weeks paid holiday every year. They work a 30-hour week. The German autoworker is the highest paid in the world. The average German autoworkers earn $39 an hour in wages and benefits. In the U.S. the average autoworker earns about $25; in Japan $27. . . . Piech has stated that some 30,000 of his German employees are redundant but he can't lay them off."
3. Robert S. McNamara with Brian VanDeMark, *In Retrospect* (New York: Random House/Times Books, 1995). McNamara makes it clear that he and the whiz kids at Ford installed cost-accounting and operations-research techniques that they were convinced made crucial differences in winning the Second World War (pp. 8–9). The public record and private conversations make it clear that he brought the same culture and values to the Department of Defense. They won one shooting war and one business war: Why not go three for three? The successful substitution of *analysis* for right strategy and insightful policy was *confirmed,* after all.
4. Author's conversation with the no longer junior executive, 1998.
5. *Forbes,* 6 January 1998, and *Business Week,* 10 December 1998.
6. "The Power of Limited Thinking: Small Scale Minds May Pay Nonrandom Dividends," *Science News,* 22 November 1997.
7. "Chimps May Put Their Own Spin on Culture," *Science News,* 12 December 1999, 342.
8. Stephen E. Ambrose, *Citizen Soldiers: The U.S. Army from the Normandy Beaches to the Bulge to the Surrender of Germany, June 7, 1944 to May 7, 1945* (New York: Touchstone Books, 1998), 441.
9. Bruce D. Berkowitz and Allan E. Goodman, *Strategic Intelligence for American National Security* (Princeton, NJ: Princeton University Press, 1991), 99–100.
10. Cialdini, *Influence—The Psychology of Persuasion,* 61–65.

11. Ibid., 117.
12. *The Oregonian*, Science section, 1 April 1998.
13. Toni Mitchell, "Big Yellow Taxi," *Ladies of the Canyon* Album, copyright 1970, Siquomb Publishing Corp.
14. Cialdini, *Influence—The Psychology of Persuasion*, 290.
15. Both the company name and the industry are disguised.
16. "Doctors' Orders: How Columbia/HCA Changed Health Care, for Better or Worse," *Wall Street Journal*, 1 August 1997; "Just What Crime Did Columbia/HCA Commit? *Wall Street Journal*, 20 August 1997; "Columbia/HCA's Internal Probe Finds Apparent Wrong-Doing in Some Markets," *Wall Street Journal*, 20 August 1997.
17. "Columbia/HCA Sues Former Executive, Alleging a 'Pattern of Racketeering,' " *Wall Street Journal*, 16 October 1998.
18. *New York Times* and *Wall Street Journal*, 5 January 1998.
19. *Wall Street Journal*, 5 January 1998, A8.
20. Ibid.
21. *Business Week*, 21 December 1998, 98.
22. Ibid., 97–98.
23. Tony Buzan, *Use Both Sides of Your Brain* (New York: E. P. Dutton, 1983), 86–115.

Chapter Ten

Top Management Teamwork: Tools for Harmonizing Strategy and Culture

Almost always the creative, dedicated minority has made the world better.

—MARTIN LUTHER KING, JR.

Managers must manage.

—PETER DRUCKER

The will to manage is essential to full success in any kind of business. . . . What Mr. Sloan brought to GM was *the will to manage.*

—MARVIN BOWER

We don't look for the best people. We look for the best team.

—FRANK SHRONTZ, CEO, BOEING AIRCRAFT, 1988

A special group of people within the corporation is ultimately responsible for the hard-driving implementation of strategy and the molding of culture to support strategy: the top management team (TMT). The TMT is the vital bridge between a company's culture and its strategic vision, whether that team is running a huge, diversified conglomerate from an office tower in London or a small, intensely focused business unit operating with two phone lines in a converted warehouse in the American Midwest.

So much is obvious. What's less obvious, perhaps, is that the OCE vision is particularly demanding on top management. Just look at what OCE demands: The OCE vision demands continuing experimentation, persistent revision of strategic insight, and constant surveillance and adjustment of resources among Hammer and Pivot, offense and defense, breakthrough, exploitation, consolidation, and harvest.

It is not enough to assemble a top-flight group of managers and expect, because they are the best, that they will achieve the best. Under OCE—and under the stress of today's ever-shifting market and competitive environment—line and staff roles shift rapidly, company culture and subcultures must be prodded to evolve in accordance with strategy, and there is ambiguity in the results of strategic tests and experiments. So, far from the current mantra to "give our great people autonomy to do the job as they see fit," today's business environment in general and OCE in particular *require intimate, sustained, mutual involvement* of top management team members. They have to be good neighbors without good fences.

Unfortunately, these stresses and strains, while requiring greater teamwork than ever, actually exacerbate tendencies to look out for number one instead. What will my job be in the next year? The next quarter? What counts as success? Will I be playing defense or offense? Will I be part and parcel of an effort at strategic expansion or consolidation? All these uncertainties that crowd the minds of TMT members corrode the team's ability to speak with a single voice and act in effective concert. Yet that is the very thing the new business environment requires. The new environment, it seems, is almost devilishly designed to focus every manager on *individual* survival and success. That sucking sound you hear is the new business environment vacuuming team cohesion out of the soul of the corporation.

There is a term for the glue that makes top managers work well together: *team cohesiveness*. Great teamwork can be considered the crucial output of top management cohesiveness in the face of all the forces that tend to rip the team apart. To implement strategy, especially in the face of an indifferent, uncomprehending, or hostile culture, it's particularly important that the TMT members be cohesive among themselves. If the TMT is riddled with politics, dissension, or jealousy once a decision has been made, or if the agenda of one senior manager isn't fully congruent with those of the others or with the strategy as a whole, it is unlikely that the rest of the company will be able to put the chosen strategy into effect. The managers won't be able to manage.

This chapter is about *tools to create TMT cohesion*. The goal of creating TMT *cohesion means amplifying the centripetal force of the personal bonds among team members so that it is far stronger than the centrifugal forces of passion and self-interest*. TMT can thus be likened to a solar system, where the gravitational attraction of the strategic sun counteracts the inherent tendency of the planets to hurtle off in their own directions.

This cohesion is the necessary condition for the effective work of teams—teamwork. Although much has been written about teamwork and how to get it, insufficient attention has been focused on its primary ingredient, cohesion.

Mechanisms for Creating Top Management Cohesion

Let's get one thing straight at the outset. To create team cohesion, we're not talking Outward Bound, river-rafting trips, or T-groups here. Although death-defying mountain climbs and wilderness treks purchased as canned interventions may be fun and may have their place, experience suggests that the lift they provide to teamwork is often a temporary illusion. One superb former regional vice president of a large medical supplies distribution company told me about such a trip: "I overcame my personal fear. They overcame their personal fear. But literally two weeks later it was all the same back at the office. Because the situation hadn't changed, the incentives hadn't changed, and the people hadn't changed."

So I don't believe that any set of mechanisms can produce TMT cohesion "on demand." It's the old "royal road" fallacy rearing its disfigured head once again. Instead, we're talking about new (and age-old) analytics, systems, skills, and structures that companies can discover, develop, and implement.

If we distill the historical experience of religious orders, sports teams, military units, and high-performing business units, we find that team cohesion can be defined as the output of a set of initial conditions:

1. Making the fate of the individual the fate of the unit.
2. Giving the team common goals that are concrete and tough.
3. Ensuring a high frequency of association.
4. Ruthlessly weeding out untrustworthy team members.
5. Creating exit barriers to leaving the group.
6. Making the team as small as possible.
7. Choosing generalists for top team positions.

I am not saying that these particular requirements are *sufficient* to create teamwork. A mission of overriding importance that fires the imagination, competent leadership, the resources to accomplish the mission, and self-belief in the ability to accomplish the mission are also required. I have simply selected the elements that seem to be extremely helpful for creating team cohesion, teamwork's vital ingredient.

Making the Fate of the Individual the Fate of the Team

It has long been noted that there is a major difference between trying to achieve stardom and trying to achieve team success. We see examples on sport teams, when an individual sacrifices the success of the team by attempting to spotlight his or her own actions. American society cheers individual achievement. America venerates the innovative or successful loner. But team cohesion depends on a bonding of individuals, which sustains their commitment to achieving common goals. When the group is rewarded for achieving common goals, actions on behalf of the group are reinforced, and commitment is built for the future.

By contrast, consider the signal a CEO gives if the TMT does poorly but individuals are rewarded or recognized. The signal is that it's every man or woman for him- or herself. Although that may be "fair" in some sense of the term, it encourages TMT members to try to achieve stardom rather than team success. That's what Frank Shrontz meant when, in choosing the people to develop Boeing's next generation of aircraft, he said, "We don't look for the best people. We look for the best team." The market research by vice president of marketing may be brilliant. But if it doesn't match what the sales department can deliver, or if the marketing chief won't translate his data into brochures, presentations, and conference call exhibits that the vice president of sales can use, revenues might not budge next year. Rewarding the head of marketing for brilliance and penalizing the head of sales would quickly split the team.

This implies another team-building policy. Someone might say: If the group is rewarded rather than individuals, and the strategy stalls out, aren't the strong players likely to quit? Answer: Not if the weak players are replaced.

All this suggests redesigning bonus compensation systems to promote cohesion. The typical TMT bonus package consists of three parts: (1) a bonus based on meeting the specific objectives for that manager, (2) a "judgmental" award that is totally subjective, and (3) a bonus tied to overall profitability.

Only the overall profitability bonus promotes cohesion: Profitability is the result of team effort. This typical package also ignores the need to change performance rewards depending on what stage of the OCE cycle the company is in. Profitability should be the measure for the bonus only if the strategy in place is in the latter stages of the OCE cycle. Those are the stages in which the strategy is being harvested for profits, and it is often the easiest part of the cycle to manage.

On the other hand, at the beginning of the OCE cycle, the TMT's bonus should relate directly to the achievement of the overall goals of the strategy. Typically, this means the identification of a breakthrough and the rapid, competent exploitation of that breakthrough.

Success may well be measured in units far different from profit dollars. It could be the number of new accounts captured, the setting up of a Web site with an ever-increasing number of daily hits, or the successful negotiation of a crucial joint venture. It could include measures of cost-effectiveness and accounts not lost when a big part of the strategy is shifting portions of the company to defense.

The idea that profit may not be the appropriate criterion for TMT reward casts a shadow over other popular monetary measures: stock price, return on capital, value added, and so forth. It all depends on where the company is in the execution of its strategy—where it is in the OCE cycle.

So far, we've talked about the pluses and minuses of profit as a measure of team success. That is easy to understand. However, the results that need rewarding and the results that can befall a strategy can be other than monetary success or failure. They can be fame or infamy within the industry or the broader public; they can be promotion to wider responsibility; they can be the opportunity to roll the entrepreneurial dice once again. The important thing is that the money, fame, promotion, and roll of the dice accrue to the TMT as a whole.

So, if a major obstacle to strategy implementation is that your company's TMT isn't pulling together, the bonus plan is a place to look.

Giving the Team Common Goals that Are Concrete and Tough, but Doable

It's well known that cohesion often blossoms when a team is required to share common goals and things are tough. In fact, under circumstances of physical hardship, cohesion often occurs spontaneously, without the intervention of designated leadership. The Boy Scouts and Outward Bound take advantage of this. (Unfortunately, in the business world, cohesion usually dissipates once the lid of physical hardship is taken off the team-building pot and team members go back to their office cubicles.)

If cohesion can blossom so easily under conditions of hardship, why don't teams pull together when they're asked to face business difficulties? The main reason is that corporate goals are often remote and intangible compared with physical hardships. Even potential bankruptcy doesn't have the urgency of an upcoming thunderstorm on a nearby ridge. And because they're remote or intangible, there is often disagreement about who should do what about the problem. Thus, corporate crises often don't compel common effort.

So the first step is to set goals and working conditions that require the sharing of common difficulties, hardships (mental, managerial, and physical), and deadlines that can serve to bring team members together.

Second is to limit goals in the beginning to extremely nearby and local ones, so that philosophical disagreements about strategy or styles of management are submerged to immediate tasks.

The third step is to repeat as a drumbeat this kind of goal setting in order for team members to learn how each of them acts and reacts to one another, to assignments, and to problems. This learning about other team members is the basis of trust, and without trust, there is no TMT cohesion.

The idea of trust is so important to TMT cohesion that it is worth a small digression. It differs radically from the ordinary idea of trust as an emotion. I see *trust as a cognitive accomplishment.* Although we often trust people we like, sometimes we like people even though we can't trust them to do the right thing. And the reverse is true. We can often trust someone we don't like, because we can depend on him or her for certain reactions. Trust is based primarily on *experience of the predictability of a person's actions.* And this pattern of predictability can happen only after repeated observations of a person's actions in real-world—not artificial—circumstances. (Charisma can be defined, then, as the ability to inspire trust without the accompanying observations.)

The fourth step in building team cohesion is to translate longer-term, abstract goals into immediate observable ones and make the common dangers evident to all team members: "You can't fail in this because. . . ." This, too, helps create the conditions under which trust can develop. After trust has been built, the goals can be made more abstract—but never more than is necessary.

For example, reaching a profit or cost reduction goal is often accomplished by achieving specific intermediate objectives. These objectives may be just the tangible goals needed for the team. To create trust and then cohesion, it is better to focus the team on the successful installation of the cost-cutting computer system, the zero-defects program, or the capture of specific customers rather than the abstract goal of bottom-line results or even 10 percent overall cost reduction.

Why does this work? First, of course, reaching agreement about something concrete is often easier than reaching agreement on abstract goals. Often, TMT members disagree philosophically but can agree on short-term objectives. At one distribution company I served, TMT members disagreed vociferously about whether the company should gradually change its whole product line mix, but they agreed unanimously on the need to improve customer service. At least that got done immediately, and it helped the product mix problem, too.

Equally important, concrete, immediate objectives help people structure work and time: who is doing what, when. Structured tasks and structured time provide a secure feeling. And since humans are territorial animals, work that is structured also helps define relationships and improve harmony.

Finally, it is useful to see trust as a kind of bank account. One person's trust in another is built up of a number of predicted observations of behavior that have been borne out. (In common speech, this is even confirmed in the negative: "Yes, you could trust old Joe to steal a dollar if it was left on a table.") A period of structured work built around near-term tasks allows team members to build up trust in each other's bank accounts.

Later, the reservoir of trust that's been created through these short-term efforts can carry over when the goal is large, remote, or intangible. For example, the corporate goal might be the successful establishment of a foreign subsidiary. Because of the shared experience and the trust that has been created in personal, day-to-day contact, the team members trust and support the task force that has been sent abroad, even though to them, the goal is remote and the people are far away.

One more thing needs to be said about goal setting. Recent research shows that people experience "flow"—a feeling of well-being, closeness, and complete identification with their work—when the goals set are not easy or trivial but demanding—but not so demanding as to seem impossible. Top sports coaches do this by continually asking for improvements that are just a tiny bit better than an athlete's personal best. The *stretch target*, not asking for the moon, is the best way to get top performance.

So, overall, my suggestion is that the chief executive in charge of strategy implementation create repeated, concrete goals that the TMT members must struggle together to achieve but that are eminently achievable.

Ensuring a High Frequency of Association

If trust is one of the conditions for TMT cohesion, one of the main building blocks of trust is a high frequency of association among team members. Again, many organizations already know this: It's not just parsimony that makes the Boy Scouts, the Army, and IBM sales trainees share living quarters.

If it's true for the rank and file, it's also true for senior managers—perhaps even more so. Top managers are often recruited from the four corners of the earth; they come packaged with individualism, competitiveness, energy, and a determination to make their own way in the world. Their operations are often geographically widespread, and their departments are often so large that they can take up floors and floors of office space, and sometimes whole buildings. So companies must often make a special effort to create the condition of high-frequency association.

Not long ago, I worked with a financial services company whose expansion required a move to a new building. For the first time, the CEO, president, and senior vice presidents occupied offices on different floors. In a flash, communication and mutual understanding deteriorated markedly. The CEO had to revamp top management's weekly meetings to ensure through formal means what had once been accomplished through face-to-face communications.

A smart CEO avoids geographic dispersion whenever possible and ameliorates it when it is unavoidable. Companies with top executives scattered across the continents need to create occasions to reduce geographic separation among TMT members. Here are some ideas I have seen work to increase association:

- Hold weekly teleconference calls that are mandatory and bad form to miss. Attendance in person or by conference call is required, with a stringent hurdle for excuses.
- Provide offices at the corporate center for regional vice presidents and hold rotating monthly management meetings throughout the country (or around the globe).
- If rotating monthly meetings are impractical, invest in video conferencing capabilities. A $1 billion company I worked with used this technology to successfully integrate its brand-new Irish acquisition.
- Consider establishing a temporary (or semipermanent) central meeting location. One company I worked with, headquartered in Los Angeles but with operations located nationwide, rented space in a suburban office building near Dallas–Fort Worth Airport. Executives based on both coasts could make it to Dallas and back on the same day as the meeting. Alternatively, it was often easy to schedule cross-country trips via Dallas to attend the monthly meetings.

Substituting lengthy off-site retreats for frequent interaction doesn't work. For one thing, it's not the total length of time spent in close association that builds trust but the frequency of association. A week of concentrated team-building sessions or "quality time" won't make up for a year of isolation. For another thing, the special environments and special tasks found in retreats breed special behaviors. What's actually needed is frequent association under normal business pressures. Retreats can rejuvenate existing teams and help TMT members get to know one another on new levels, but they can rarely provide the bedrock for team cohesion.

How important is frequency of association? In the German army during World War II, recruits in every unit came from the same region of the country. Also, replacements for casualties arrived at the front in replacement battalions, where they were trained by the "very officers and NCOs who would later command the men in battle . . . the replacements would reach the front already knowing both each other and their commanders and forming part of a well-integrated team."[1]

By contrast, America went a bit too far in its love affair with the assembly-line model. The most striking example was the American scheme for assigning brand-new replacements to front-line units. First, upon recruitment, they were processed using standardized tests. This (and manpower needs) slotted them for what's now called an occupational specialty. Once they graduated from stateside basic training, American GIs were treated as interchangeable cogs for the front-line fighting machine. Saying good-bye to their basic-training buddies, they traveled as lone individuals from basic-training camps through various staging depots to the front line, a trip that normally took four to five months. They arrived "tired, bewildered and disheartened. . . . Field training bivouacs usually were within the sound of guns, and the replacements were acutely and nervously aware that their entry into combat was imminent. They frequently did not know how to take care of themselves."[2] The highest casualty rates occurred among these newcomers, often to the point where the front-line veterans avoided even learning their names, much less teaching them how to survive. In sum, "the U.S. Army by contrast [with the Germans] put technical and administrative efficiency at the head of its list of priorities, disregarded other considerations, and produced a system that possessed a strong inherent tendency to turn men into nervous wrecks. Perhaps more than any other single factor, it was this system that was responsible for the weaknesses displayed by the U.S. Army during World War II."[3]

The results were murderous. According to historian Martin van Crevald, on average, and taking into account differences in equipment and the inherent advantage of the defense, the German army inflicted 50 percent more casualties on the Allies than it suffered. "This was true when they were attacking and when they were defending, when they had a local numerical superiority and when, as was usually the case, they were outnumbered, when they had air superiority and when they did not, when they won and when they lost."[4]

Ruthlessly Weeding Out Untrustworthy Team Members

Getting rid of untrustworthy team members may be the single most important action a CEO can take to secure implementation of a corporate strategy. Without this action, none of the other actions are worth the effort. It is also one of the most seldom undertaken, and one of the most broadly needed.

Two cases of my own experience illustrate both the necessity and the difficulty of weeding out untrustworthy TMT members. A person we'll call Ron Johnson was the controller for a half-billion-dollar distribution firm. Everyone acknowledged his high competence in that function. At one point, despite a strict finance background, he successfully filled in for a departed vice president of operations.

But as long as Ron was on the TMT, nothing but dissension and rancor pervaded the team. Why? Ron made a habit of checking out the president's inclinations before any operating review meeting and then supporting those inclinations regardless of what he might have said in the past and regardless of the facts. Surely you know the type. Further, as the single TMT member with access to the details on financial statements, he almost always provided the numbers that would support his viewpoint du jour. After six months or so, he was perceived as a completely cynical Byzantine courtier. The rest of the TMT began to keep quiet in regular management meetings. When they solved problems together, they worked them out in nearby bars, without the benefit of the numbers.

The president himself was far from ignorant of the situation, although I believe that he grossly underestimated Ron's perfidy. He underestimated even more the damage to the TMT. He believed that he could use this successful but self-aggrandizing manager by giving him hard missions and carefully evaluating whatever he accomplished. In other words, he believed that, as with many other kinds of human flaws, this too could be worked around.

The tough assignment approach just didn't work. Ron got the job done, but only through brute force and without the consent, conviction, or cooperation of the other TMT members. Although the good-natured president tried to filter the facts and information the controller supplied him, he had no source of contrasting facts with which to sort wheat from chaff. And he had other small matters on his calendar, such as keeping major vendors on his line card and large customers happy.

In the end, Ron realized that he would not be awarded the promotion to CFO he'd hoped for. He left for an aluminum company. After his departure and that of two other less-than-competent or untrustworthy TMT members, the remaining close-knit group pulled together. They succeeded in moving the company back from the edge of bankruptcy and improving its performance enough to achieve its sale at an unexpectedly high premium.

My second story involves a large hospital chain. The director of purchasing had achieved widely acknowledged success in negotiating lower prices for supplies. However, he had alienated the field TMT managers by his authoritarian manner and by "slow-rolling" the changes they requested. In public, and to their faces, he gave every appearance of supporting them. But by demanding more information and time for further study, he killed most requests through maleficent neglect.

Not surprisingly, the field managers recommended a drastic slashing of central purchasing's budget when the time came for severe cuts in corporate overhead. In the meantime, the CEO and president decided that to gain field cooperation on supply and logistics issues, the purchasing director would have to be dropped from the TMT and a new director of logistics inserted over his head.

There are four common difficulties in firing untrustworthy TMT members: First, they are often very competent—and appear competent—within their functions. They may be hard to replace, for they are often clever enough to make themselves indispensable.

Second, they can be hard for their bosses to identify: They often worship at the temple of camouflage. In the case of Ron Johnson, although the president had knowledge of the problem, he had no way of knowing how bad it was.

Third, for a CEO trying to get something done, it's often a case of a known evil being better than an unknown one. If I fire this person, who's the replacement? How long will it take to get him or her up to speed? How do I know that the replacement will be any better? There are always reasons for postponing the unpleasant.

Fourth, untrustworthy team members are often clever. They try to leave no grounds for dismissal "for cause" and often have negotiated tight employment agreements. They know well the game they are playing.

But the consequences of *not* removing untrustworthy team members are extremely high. The TMT will never work smoothly, and it is a rare strategy that is robust enough to withstand implementation that is hobbled by distrust from the start.

The best way to tackle this unpleasant task is to plan it like any other project. Aside from the legal and personnel conditions you need to fulfill, the all-important, long lead-time step is to cultivate a backup person or team over a period of weeks. And the key action here is to grab hold of the information and relationships the targeted TMT member has at his or her disposal. You can request special reports and analyses in the course of business to gather the information. Relationships are trickier, but ride-alongs to vendors or customers are routine enough to provide a reasonable pretext—not to mention a way of testing the veracity of any rumors about untrustworthiness.

Creating Exit Barriers to Leaving the Group and Purging Anonymity

A beautiful strategy can be difficult to implement, and it can take years to bear fruit. So it often happens that the lifetime of the OCE cycle is longer than the average tenure of a TMT manager. It is simply too much to demand of human nature to expect those who have easier or more lucrative opportunities elsewhere, or those who are nearing retirement, to bend wearily at the oars of strategy implementation when the harvest of their efforts is beyond their personal time horizons. That is why barriers to easy exit—not without compensating rewards, of course—are a vital condition for creating TMT cohesion. "We're in this together" is a powerful realization.

Some widespread management practices not only leave the exit doors unlocked but also employ a concierge to open the doors. Golden parachutes—which are supposed to open only in the event of a takeover or merger—often encourage that very result. At a large medical company I know, more than one vice president told me that he wished management's organizational overhaul program would fail quickly. The inevitable change of ownership would make them rich.

Another widespread practice is stock options for companies expecting to go public. Senior TMT managers at a financial services start-up I've worked with spent as much time counting the increase in their net worth upon the expected initial public offering as making the strategy work that the IPO was predicated on. Several had plans to leave the company shortly after the IPO if things weren't going to their satisfaction.

The current and projected shortage of top management talent in the early part of the new millennium can only exacerbate turnover in the TMT. However, there are some actions a CEO can take to cut back on the temptations.

On the compensation front, instead of creating wealth for TMT members at a single stroke, install systems that provide much greater individual wealth over a much longer period. So, for example, instead of a nice stock option "pop" upon a company's going public or reaching a certain stock market value, make the "pop" much bigger but much longer term. This is done through managing vesting requirements and matching rewards to the achievement of long-term strategic goals.

Probably the biggest exit barrier you can erect, however, is non-monetary. (This is a good thing, because just as there is someone out there who is younger, smarter, or better looking than you, there is nearly always someone who can outbid you.) Instead of money, it is reputation. Throughout history, organizations—from centuries-old

British regiments to sports clubs to twelve-step recovery programs to multilevel marketing sales forces—have exploited the human need for widespread recognition by others. Often, reputation can be substituted for monetary compensation, because once people achieve a certain level of comfort, what they want more than anything else is the esteem of their fellow human beings. Yet except for high-visibility CEOs, many effective managers at the senior executive ranks remain almost anonymous in large companies and small.

The solution, then, is to purge the TMT of any lurking anonymity. Instead, link each TMT member's public identity to strategy-supporting activities. For example, suppose you are making the rounds of Wall Street to raise equity for your start-up. You've hired a spectacular vice president of sales. Have her make the presentation, and thus the public commitment, instead of letting the president or CEO do it, and make her available to answer questions. The more identified she becomes with the fate of the strategy, the more difficult it will be for her to quit without having succeeded. What would she say to those investors the next time she met them?

Other venues that offer opportunities to link TMT members publicly to portions of their strategies include industry meetings, trade publications, and even internal meetings of TMT members and the company's workers. The rule is to take advantage of every opportunity to purge anonymity from strategy implementation efforts.

One of Napoleon's greatest leadership tricks was to make unannounced visits to his troops and hand out surprise medals and rewards that made the favored soldier's companions gasp with envy. He realized that a little esteem can make hardship, poverty, and even wounds and death seem worthwhile.

Making the Top Management Team as Small as Possible

One of the quandaries of strategic leadership is that top management is always shorthanded. But adding members to the TMT often seems to result in getting less done rather than more. And it dilutes the chances to boost team cohesiveness.

The purpose of this section is to suggest that top managers responsible for strategic implementation resist the impulse to include any people in the TMT. This impulse seems to be inherent, simply because there is so much work to do. For example, during a four-month consulting assignment for a start-up financial services company, the number of invitees to the weekly management meeting grew (in disguised form) from:

1. CEO
2. President (also responsible for national account sales)

3. Senior vice president (SVP) quality and operations
4. SVP ABC services (a product line)
5. SVP DEF services (another product line)
6. Vice president (VP) controller
7. VP customer service
8. VP loan processing
9. VP information services
10. VP human resources

to:

1. CEO
2. President
3. SVP quality and operations
4. SVP marketing
5. Chief financial officer
6. Controller
7. SVP ABC services
8. SVP DEF services
9. SVP information technology
10. VP controller
11. VP customer service
12. VP loan processing
13. VP information services
14. VP purchasing
15. Corporate counsel
16. VP human resources

That's a 60 percent increase in less than a third of a year. Of course, for every one of these additions there was a perfectly valid reason. Just as surely, no one was dropped from the group. But even the main boardroom was getting too small for this, the company's main strategic planning and implementation meeting. And I know that this company is far from alone. If the TMT recipe includes such functions as government relations, regional and international vice presidents, research and development, business development, and human resources, and you divide sales and marketing in two, and then it's flavored with a few special assistants, you can easily reach a TMT of twenty to twenty-five members.

And as an observer to the more modest transformation mentioned above, I found it hard to believe that decisions were made either faster or better or implemented more completely.

One reason is that the math is against you. First, let's focus on one-on-one communications. Let's assume that it takes each TMT member an average of two hours a week to keep current with each other member. Let's also assume that each TMT member works sixty hours a week. Thus:

- A 5-member team would consume 13 percent of its time on one-on-one communications in a 60-hour week (5 members × 4 meetings × 2 hours = 13 percent of 300 available work hours).
- A 10-person team would consume 30 percent of its time on one-on-one communications.
- A 15-person team would consume 47 percent of its time on one-on-one communications.

See Figure 10-1 for a graphic representation of this inherent inefficiency.

Thus, with a fifteen-member team, almost half of available work time is spent merely keeping up with what other team members are doing. What's important to notice is that by tripling TMT size and thus hoping to triple the number of management hours available to implement strategy, we have actually increased the time available from

Figure 10-1. TMT size decreases percent of time available for productive work.

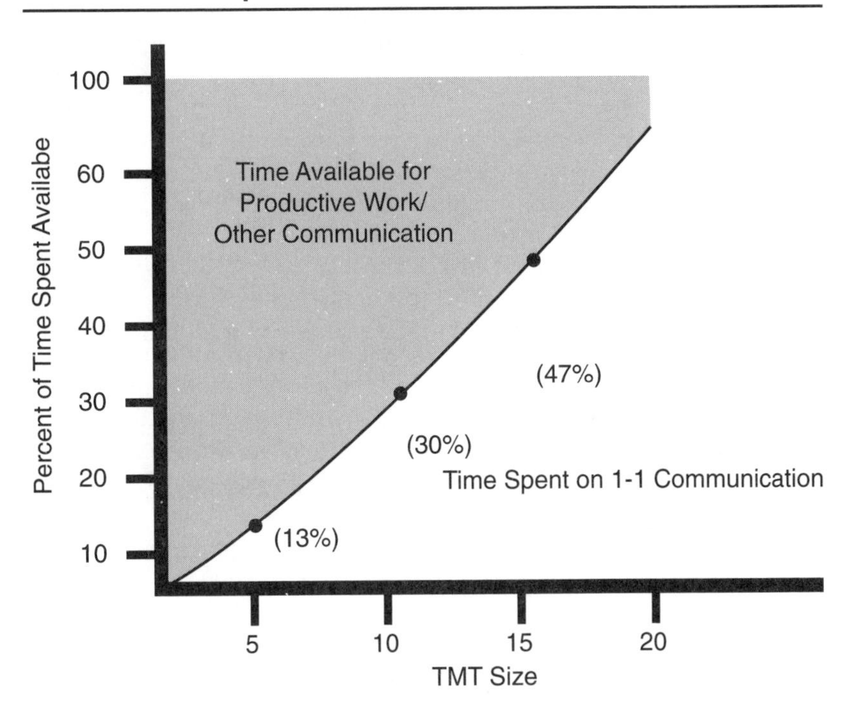

260 hours per week to 480—less than 100 percent for a 200 percent increase in staffing. (Here I define time available to implement strategy as total TMT time available minus the time needed for one-on-one meetings.)

The math is also against us in another way. As the number of TMT members increases, the number of one-on-one relationships within the TMT increases exponentially according to the formula $n(n-1)/2$.

- For a 5-member team, there are 10 one-on-one relationships.
- For a 7-member team, there are 21 relationships.
- For a 10-member team, there are 45 relationships.
- For a 17-member team, there are 136 relationships.

That's a lot of relationships that have to be nurtured. And you can see how they grow much faster than the number of team members. This is the mathematical basis for corporate policies.

We also know intuitively that the larger a team is, the more difficult it is for the CEO to get it to implement a decision with all the functions fully coordinated and pulling in the same direction. However, this difficulty increases exponentially, not arithmetically, with the number of team members.

Again, to oversimplify, let's assume that each TMT member's position on a strategic issue could be characterized (from the CEO's point of view) as "correctly understanding" or "not correctly understanding" the issue. Let's further assume that for every combination of ways the TMT could be arranged, the company would head in slightly—or greatly—varying directions, since each function represented on the TMT would be pulling the company in the direction of the understanding of its TMT representative.

Now, where n is the number of TMT members, there are 2^n number of ways the TMT can be arrayed on any issue. And so, according to our simplifying hypothesis, there can be just as many different directions the company might be pulled in:

- For a TMT of three people, there are eight combinations possible.
- For a TMT of five people, there are thirty-two combinations possible.
- For a TMT of ten people, there are 1,024 combinations possible.
- For a TMT of fifteen people, there are 32,768 combinations possible.

See Figure 10-2 to understand how fast misunderstandings can spread.

Figure 10-2. Beyond 10 TMT members, possible misunderstandings increase beyond top management's ability to control them.

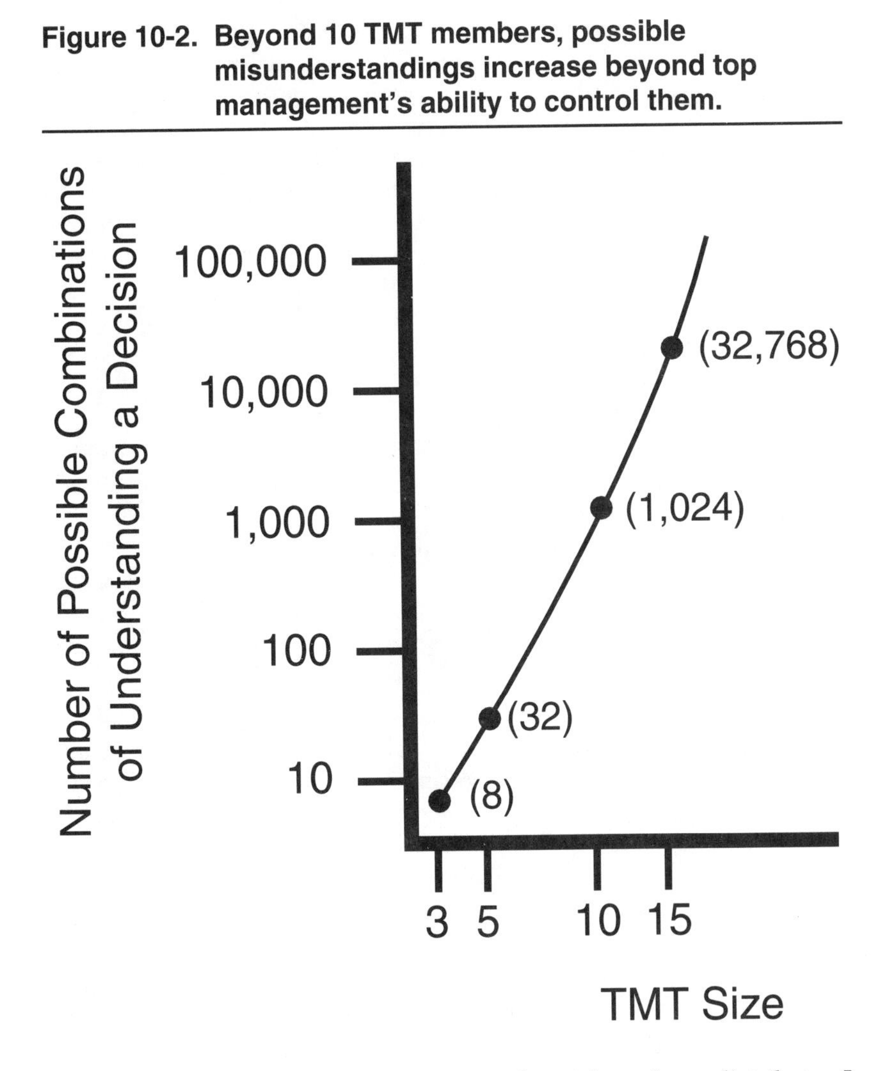

Take the case, for example, of an industrial products distributor I worked with. The TMT consisted of the new president, four regional vice presidents, the VP of sales, and the controller—a small team. But the company was in grave danger. Its two largest suppliers had terminated their contracts, accounting for over 20 percent of sales. More important, these were such leading suppliers that the TMT feared that customers would cross the company off their "preferred vendor" lists. It was no wonder that the company's bankers were beating the door down when the president announced a turnaround plan that focused on customer service. But the company couldn't get any traction to implement this plan. Even this small team was up against the mathematics of misunderstanding: Even if there were only two ways for each TMT member to understand what "improved customer service" meant, there was (mathematically) a possibility of 128 different implementations of this policy—not a fabulous way to create team cohesion.

In reality, there were perhaps half a dozen ways to understand the customer service concept. Did it mean answering the phones more quickly and getting bids out more quickly? Reducing order fulfillment error rates? Bending over backward to meet customers' (often unreasonable) shipment deadlines? Did it mean improved sales force call rates and better inside sales? Once the TMT realized that it was getting no traction on the turnaround plan, it took three full days of meetings to hammer out exactly what the president meant and what *that* meant for each TMT member.

To counteract these tendencies, CEOs face some awkward alternatives:

- They can put up with the inefficiency and insist on a high level of communication among team members. This appears to happen in "introspective" companies such as Xerox in the late 1970s and early 1980s and Tektronics. But it is unworkable when the market situation demands a rapid tempo of decision making.

- They can put a lid on the amount of internal TMT contact and communications by cutting back on who is invited to meetings and by making demands for TMT productivity. In this case, contact and communications are highly intense within related subgroups of the TMT (for example, the CFO, chief counsel, and controller), but communication across these groups withers.

- They can reduce TMT size to a minimum, giving each TMT member broader responsibilities. There is evidence that this can work: Charles Wang of Computer Associates, a large, highly profitable software development company, speeds up projects that are behind schedule by cutting, not adding, programmers. "Take your two weakest ones off. It's a very simple process and it works every time."[5]

So what *is* the right size team? Perhaps there is no definite, one-size-fits-all answer. But experience suggests an amazing consensus around the number ten. Genghis Khan's army, the Roman legions, the British army, the U.S. Army, and others are based on squads of ten to eleven. Colonel William Darryl Henderson states that cohesion is inversely proportional to the numbers in a group. "Several armies . . . have determined that the ideal size is up to nine men, with some armies choosing a three-man unit or military cell . . . [as] the basic . . . building block."[6]

Jaundiced corporate observer Anthony Jay notes that "basketball has five men and ice hockey has six, but almost all other team games have teams of between eight and 15."[7] Peter Parkinson, of Parkinson's law fame, states that "no board, council, cabinet or committee can ever function when membership rises above twenty. It will, of course, continue to meet . . . but a smaller group will have formed which does the real work and makes the real decisions."[8]

This has been my experience as well. In the end, a small "in crowd" of manageable size forms, often without anyone, including its members, being aware of what is happening. At a multibillion-dollar health care company undergoing half a dozen major change programs, the chairman found more management success by informally, perhaps unconsciously, confining the TMT to three people: himself (with a legal background), a general counsel–turned–operations vice president, and an outside strategy consultant. The president, regional vice president, and functional vice presidents still had their operational and functional rules, but each basically reports to one of these three. The group made most of its strategy and strategic implementation decisions during hour-long walks in a nearby public garden.

Choosing Generalists for Top Management Team Positions

If the TMT is going to be small, this means *not* hiring the best man or woman for each TMT job. Rather, the objective should be to hire a few people of wide experience and capability and fairly deep experience in one or two areas. In other words, *for TMT assignments,* when it comes to choosing between a person with spectacular qualifications in a narrow area and a person with good qualifications in many areas, on balance, the nod should be given to the generalist.

This is best understood by example: the company mentioned earlier with a rapidly expanding TMT. One reason for the expansion was that the generalist CEO inherited team members who were very good at their chosen specialties. But almost none of them had experience with or even much empathy for issues outside those areas. Further, investors were pressing the company to hire well-known names for crucial marketing and information technology positions, so that Securities and Exchange filings for its IPO would be unmistakably star-studded. So the TMT expands as each functional slot is filled with a terrific specialist who can handle only a narrow set of responsibilities.

This company has temporarily solved the problem of coordinating all these deep specialists, partly through ingenuity and partly through good luck. It is fortunate in having three people of highly varied backgrounds and remarkable intelligence at the helm. The CEO is a generalist with military, retail, consulting, and financial service experience. He is far smarter than any company has the right to expect in the CEO lottery. The president has deep industry experience but a notably open mind gained through travel, expatriate status, and a powerful imagination. Finally, the person who is the day-to-day glue of the company has entrepreneurial experience running a newspaper chain, software and electronics companies, a highly successful real estate brokerage company, a political campaign management company, and a multithousand-member affinity group. He is also endowed with tremendous physical and mental energy and a genial personality. His job: to understand everything that is going on and to act as a one-person management reserve to exploit opportunities, fix problems, and keep all the specialists singing from the same management score. He is the keeper of TMT cohesion.

Generous long-term stock options, total mutual trust, and an endless list of short-term but high-stakes goals on the IPO road promote TMT cohesiveness. The CEO, president, and CFO have very high public visibility, so if the IPO or company were to fail, the loss would wound their reputations for years—and they know it. Debate is vigorous, but cohesion is high. And a generalist is in charge.

So, though every force tried to turn into a triumph of the specialist, in the end, it is the generalists who keep a company glued together and its strategy on course.

Shaping Cohesion in the Top Management Team: Work for Tomorrow

Inventory your TMT situation and create an action plan to modify it. Ask the following:

1. Is the fate of the individual the fate of the TMT? Are all the rewards aligned with strategic implementation success?
2. Are stock options, bonuses, and golden parachutes creating cohesion or individualism? Long-term or short-term thinking?
3. What are the TMT's current goals?
4. How concrete are these goals? Can TMT members specify the exact target outcome and target date? If so, what are they? If not, what can be done to make them short term, concrete, and a stretch to achieve?
5. What is the company doing to ensure a frequency of association? In particular:

- What is the current frequency of association?
- What are the current barriers to a high frequency of association

(geographic, office location, press of departmental business)? Be specific.

- What three or four things can be done to increase frequency of association?

6. What exit barriers are there to leaving the group? Monetary? Contractual? Reputation? In particular, what can be done to link each TMT member's public reputation with all or some of the crucial strategic implementation tasks?
7. Reflecting on each individual team member, how would you rank his or her trustworthiness?
8. What are the next steps to weeding out untrustworthy team members?
9. How can we make the TMT as small as possible? In particular:

- What is the current size of the TMT?
- What percentage of TMT time is now spent just keeping abreast of one another's work?
- What percentage of time is available for value-producing work?
- Thinking *ideally*, if we had the right people as members of the TMT, how many members could we currently consolidate?
  ___________(persons) under ___________(TMT member)
  ___________(persons) under ___________(TMT member)
  ___________(persons) under ___________(TMT member)

10. How do we balance the need for specialist experience and generalist capabilities for TMT positions?

- The following TMT members are specialists: ___________
- The following TMT members are generalists: ___________
- The following TMT members have the capability of growing in specialized experience or becoming generalists: ___________

____

11. Overall, our action plan to improve the balance and effectiveness of the TMT is as follows:

- Over the next four to six weeks: ___________
- Over the next two to three months: ___________
- Over the next six to twelve months: ___________

Notes

1. Martin van Crevald, *Fighting Power* (Westport, CT: Greenwood Press, 1982), 76.

2. Ibid., 77.
3. Ibid., 79.
4. Ibid., 80.
5. *Forbes*, 11 July 1988, 119.
6. William Darryl Henderson, *The Human Element in Combat* (Washington, DC: National Defense University Press, 1985), 10.
7. Anthony Jay, *Corporation Man* (New York: Random House, 1971), 35.
8. Quoted in ibid., 34.

Chapter Eleven

81 Do's and Don'ts on the Road to a Great Corporate Strategy

What you do
Still betters what is done.
—WILLIAM SHAKESPEARE, *A WINTER'S TALE*

1. Don't fall in love with a strategic hypothesis, especially the current course you're on.
2. Do be suspicious of "overwhelming evidence" in favor of a strategic hypothesis.
3. Do demand that your proposed strategies prove themselves in the fire of falsification.
4. Don't seek the one right, great idea for your company.
5. Do probe for many great ideas.
6. Don't overestimate the value of consistency in your own and your staff's thinking.
7. Don't make measures of internal consistency the "gold standard" in the testing of strategic ideas.
8. Do frown suspiciously at analyses, presentations, and staff positions that are highly internally consistent—the evidence was likely forced to "fit."
9. Don't mistake a mission or goal statement for a strategy: They are ships without rudders.
10. Do make explicit and precise the conditions under which you expect certain strategic results: Work out the *if-then* statement that is your strategic hypothesis.
11. Do make explicit and precise the results you expect from the conditions you set up and set out the laboratory or field experiments to test them.
12. Do remember that it's only human nature to *imagine* that what

we see around us are the true facts of the case, including ones about corporate strategy and making money.

13. Do generate lots of strategic hypotheses—to generate a good hypothesis, it's best to generate lots of hypotheses.
14. Do ensure that you have a broad range of hypotheses, not just variations on a single theme.
15. Don't settle for broad top management consensus on a strategy before you've done your level best to falsify it: Cultures wear blinders.
16. Don't delegate hypothesis generation: It's a top, middle, and bottom management responsibility.
17. Don't procure your strategic ideas from the same old sources: yourself, consultants, marketing, strategic planning, the sales force.
18. Do seek out ideas from everywhere in and out of the organization.
19. Do use stretch targets to push your strategic thinking: Never allow that your staff's first, second, or third answer is good enough.
20. Do get top management out of the office and into the offices and parking lots of customers and competitors.
21. Don't expect market research to point you in the direction of the right strategy without its first being pointed in the right direction by you.
22. Do expect more productivity from market research, especially hypothesis generation, and especially from focus groups.
23. Do use quantitative market research, but primarily to test precisely formulated strategic hypotheses, not because you expect it to provide strategic insight.
24. Don't compete head to head—Hammer to Hammer—if you can avoid it.
25. Do identify the Hammer in your competitor's strategy: That's the only *way* to avoid it.
26. Do identify the Pivot and its Bearing in your strategy—that is your most vulnerable point.
27. Do determine the Hammerhead in your strategy and overresource it.
28. Do create mobile assets, mobile reserves, and financial reserves and use them to reinforce your Hammerhead or to help out if your Pivot or Bearing is under attack.
29. Don't focus strategy entirely on customers and "the market."
30. Do contemplate making your strategy competition-centric: Seek out the competition's Pivots, Bearings, and underlying strategic assumption and bore in on them.

31. Don't believe that your strategy gives you a sustainable competitive advantage (SCA)—they're getting as rare as dinosaur teeth.
32. Do consider the opportunity creation and exploitation (OCE) cycle as a dynamic alternative to sustainable competitive advantage.
33. Don't seek a strategy, exploitation style, or leadership style without considering the inherent variability in your business environment.
34. Don't let the annual budget process drive resource allocation; consider *dropping it altogether.*
35. Do drive resource allocation according to whether a department is playing offense or defense, whether it is in your strategy's Hammer or Pivot.
36. Do drive resource allocation based on where you are in the OCE cycle rather than on last year's budget or a department's annual action plan.
37. Don't think that there is only one way to create a real-world realization of a strategic idea: to "drive with concentration."
38. Do evaluate many "probing pinpricks" to find the breakthrough real-world realization of your tested strategic hypothesis.
39. Don't imagine that there is only one way to exploit a strategic breakthrough.
40. Do think through whether it is wiser to exploit a breakthrough "in depth" or "sideways" and consider that exploitation in depth often creates a strong Pivot for exploitation sideways later on.
41. Do consider the inherent variability of the environment before choosing exploitation in depth or sideways.
42. Don't do consulting company shoot-outs (just) at the beginning of a project.
43. Do use a hypothesis-testing shoot-out when hypotheses have been generated.
44. Do force stretch revenue, market share, and profit targets for hypothesis generation.
45. Don't think that the end of strategic hypothesis generation or implementation planning is the final presentation by your strategic thinkers.
46. Do think that it is the first half of a courtroom process in which the opposing side gets to try to refute the hypothesis and come up with something better.
47. Do use a courtroom process to test and falsify strategic hypotheses.

48. Don't measure the value of a proposed strategic hypothesis by the tenacity with which vocal managers maintain it; that is a beggarly kind of hypotheses testing.
49. Do organize conflict over competing hypotheses so that a better answer emerges from the arena of conflict than any of the participants had going into that arena.
50. Do remember that it's the *process* of reconciling these conflicting viewpoints that gives rise to a superior strategy than either view alone.
51. Don't rely on top management's or sales' or marketing's "best judgment" to evaluate a strategic hypothesis.
52. Do rely on fact-based techniques such as discrete choice modeling to test hypotheses.
53. Don't use focus groups for testing hypotheses: the sample is small, "group think" creates illusions, and your competition isn't present.
54. Do use focus groups as wellsprings of raw, unrefined ideas.
55. Do assess the value of electronic boardroom focus groups instead of traditional ones.
56. Don't expect extensive, unguided fact gathering to result in strategic insight or hypothesis generation.
57. Do rely on intuition, experience, and discussion to generate strategic ideas that can be rendered into precise strategic hypotheses, then tested.
58. Don't get wedded to a single target capital structure.
59. Do expect to vary your capital structure dramatically, depending on where you are in OCE cycle.
60. Do decide first what is the Pivot and what is the Hammer and then assign financial resources with frugality *or* profligacy accordingly.
61. Don't take for granted that your company's culture matches strategic needs.
62. Don't take for granted that your company's culture can't be changed.
63. Do inventory the influences of your company's history, personality, and power structure on its culture.
64. Do inventory the current conscious and unconscious components of your company's web of belief.
65. Don't overestimate your company's current diversity: There is less than you think.
66. Do think about creating a machine-tool culture.
67. Don't be caught unawares by the influence of culture—it is ubiquitous and stealthy.
68. Do look for the stealthy signs of culture, including the hidden

assumptions in casual conversations, office and shop-floor design, and company etiquette.

69. Do expect the diversity of people and their ideas to decline rather than increase with time. If you are in a position of power and you are comfortable with the "good thinking" of the people around you, watch out: You are at a point of maximum cultural danger.
70. Don't overlook the CEO factor in corporate culture: Culture tends to gravitate toward the CEO's personal style.
71. Do ask yourself, as CEO, department head, or regional or product line honcho: Would I be proud of a staff that imitated *me?*
72. Don't wait for a strategic or financial crisis to force a needed change in culture.
73. Do precipitate such a crisis, if necessary, to bring about needed change; sometimes it's the only way.
74. Don't take top management team (TMT) cohesion for granted.
75. Do create the ongoing conditions for top management cohesion, especially formulating concrete near-term goals, weeding out untrustworthy team members, and managing incentive compensation to make the goal of the individual the goal of the team and the goal of the team the goal of the strategy.
76. Don't believe in the tyranny of specialists—that a cadre of world-class specialists can be an efficient top management team.
77. Do recognize that increasing team size usually reduces its efficiency; reduce the top management team to the barest minimum of members—a target of ten is reasonable.
78. Do use experienced, wide-thinking generalists to leaven a top management team made up of high-impact generalists.
79. Do enact sunset provisions for policy governance.
80. Don't believe that there is any royal road to a great corporate strategy.
81. Do believe there are great strategies out there to be discovered.

Index

About the Author

Evan M. Dudik is president of EMD & Associates, Inc., a management consulting firm based in Vancouver, Washington *(www.emd-assoc.com)*. EMD & Associates specializes in fact-based, rapid-response strategic consulting, new market entry, and operations improvements for clients worldwide. He was formerly president of a wood products manufacturing company. He has been a consultant with McKinsey and Co., Inc., and a Washington, D.C., based project manager and lobbyist. Evan holds an M.B.A. from Harvard Business School, a Ph.D. from the University of Texas, and a B.A. from St. John's College, Annapolis, Maryland. For comments or questions on this book, he invites you to contact him at *edudik@pacifier.com*, tel. 360-604-1818, fax: 360-604-1914.